FOR THE LOVE OF SOIL

Praise "For the Love of Soil"

"A must-read for all farmers and land managers. Nicole uses a wonderful combination of personal experience, good case studies and evidence-based science to show how regenerative agriculture systems can improve farm productivity, reduce production costs, improve biodiversity, build resilience and how to substantially mitigate this existential crisis. It is truly a win, win, win solution to the multiple problems effecting today's agriculture."

-André Leu, International Director, Regeneration International, Ambassador, IFOAM - Organics International

"For years many of us involved in regenerative agriculture have been touting the soil health - plant health - animal health – human health connection but no one has tied them all together like Nicole does in "For the love of Soil"!

She shows us through her own personal experiences and those of farmers, ranchers, researchers and medical professionals all over the world that the answers lie in the soil. A very thought provoking read that is a call to action for all of us. I highly recommend "For the Love of Soil" for anyone interested in their health, their children's health and the health of our planet!"

-Gabe Brown, Browns Ranch, Nourished by Nature

"William Gibson once said that "the future is here -- it is just not evenly distributed." Nicole modestly claims that the information in the book is not new thinking, but her resynthesis of the lessons she has learned and refined in collaboration with regenerative land-managers is new and it is powerful. She lucidly shares lessons learned from the deep-topsoil futures she and her farming and ranching partners manage for and achieve. I read it with wonder, gaining something new and useful from every page.

Nicole listens deeply to the land and the people she works with. I was repeatedly struck by the humility, elegance and lightness of touch she brings to practical regenerative land management. The lifespan of paper copies of this book is likely to be short as it bounces around on farm-truck dashboards and in saddlebags, and is read, re-read, marked-up and referenced by working hands."

- *Abe Collins*, Collins Grazing, Co-Founder, Soil Carbon Coalition

"To understand the complexity and wisdom of nature is made easy and very readable by this global soil expert. With the knowledge she shares, it empowers us to act with confidence that we all can make a difference even if only by our choice of food.

-Tim LaSalle, PhD. Center for Regenerative Agriculture, California State University Chico; CEO Rodale Institute; Exec Director Savory Center for Holistic Management.

"Nicole Masters opens this wonderful missive with the statement "I am not normal!" As we experience her passionate call for action, we become increasingly grateful that this unique soul was never born to mediocrity. She is a gifted soil warrior!

In her unique, storytelling style, Nicole unveils a flawed food production system that is in desperate need of review. She chronicles the heroic work of the Regenerators, whom have accepted that challenge, and she provides a pragmatic and inspiring game plan for those contemplating that journey. A must-read for those seeking a hard science-based recipe for meaningful change. You will discover that it is truly possible to be more profitable and more sustainable, without sacrifice."

-Graeme Sait, CEO, Nutritech Solutions

"Out of sight, out of mind is no longer an excuse to ignore what goes on beneath our feet. Nicole Masters takes you on a subterranean tour of your operation, introducing you to a cast of invisible characters that determine the productivity of your farm and the healthfulness of the food you produce. Nicole writes like she speaks. In her colorful, no-nonsense style she tackles complex ideas and breaks things down without dumbing things down."

-Dave Pratt, Founder, Ranch Management Consultants

"We learn and ideas 'STICK' when attached to stories. 'For the Love of Soil' is a great and fun read, giving the earth hope for clean air, clean water and clean nutrient rich foods."

-Gary Zimmer, Author, *'Advancing Biological Farming,'* Midwestern BioAg

"This is an important book which will lead farmers and graziers, agronomists and scientists into regenerative practices and thought processes. Our global agricultural production systems are frequently at war with ecosystem health and mother nature. In this book, Nicole is declaring peace with nature and provides us with the science and guidelines to join the regenerative agriculture movement while increasing profits."

-**Terry McCosker,** Director, Resource Consulting Services, Carbon Link

"What does healthy soil look like? How do we determine the state of the soil under our care, whether it be a ranch, farm, garden or yard? And, most importantly, how do we improve its condition? Nicole Masters tackles these questions masterfully in her book For the Love of the Soil"

-**Lill Erickson,** Executive Director, Western Sustainability Exchange

"For the love of soil," captures the elegance, intricacies and power of nature! Nicole's insight and experience lays out the case for biological regeneration of this planet's soils and ecosystems.

Exemplifying a solution in the statement "it's the structural supports and thinking that has got us into the pickles we are facing in agriculture, and it will take a shift in our thinking to discover long-term solutions." This book describes and demonstrates this shift towards logical, common-sense and ecosystem-friendly practices which promote those profoundly needed solutions!

I recommend this book to all so that they can develop a better understanding of how life has shaped this earth and how we can align ourselves with these processes to fashion a better future for ourselves, our progeny and this planet!"

- **Dr David Johnson,** Adjunct Professor, California State University, Chico, Center for Regenerative Agriculture and Resilient Systems

"Nicole digs deep into the soil and fungal networks that nourish us. She understands roots, stems, leaves, worms, mycelium, sunlight, sugar, oxygen and carbon, and weaves them together in this beguiling and essential book about the practice and dirt-deep wisdom of regenerative agriculture. Essential reading if we hope to save life on the planet."

- **Gretel Ehrlich,** Author, *"The Solace of Open Spaces"*

FOR THE LOVE OF SOIL

For the Love of Soil

Strategies to Regenerate Our Food Production Systems

Nicole Masters

 This is a work of nonfiction. No names have been changed, no characters invented, no events fabricated. Although I am a coach, I am not your coach. Reading this book does not create a coach - client relationship between us. I do not and cannot assume any responsibility for property damage or loss of production as a result from any misinterpretation or misapplication of the information and suggestions in this book. All suggestions in this book should be trialed on small areas first, no outcome is guaranteed. This book should not be used as a substitute for the advice of a competent coach, consultant or agronomist.

Interior images and photography: Nicole Masters

Cover design and interior print design: Jeremy Masters

Cover conception: Bryn Masters

Author photo: Laura Nelson

Proof reading, editing errors and the use of New Zealand spellings are all the fault of the author.

First published in New Zealand in 2019 by Printable Reality:
www.printablereality.com

Library of Congress Cataloging-in-Publication Data

Masters, Nicole R.

For the love of soil/ Strategies to Regenerate Our Food Production Systems [by Nicole Masters}.

p.cm

Includes bibliographical references and index.

ISBN 978-0-578-53672-9 (pbk)

1.Sustainable Agriculture 2. Soil Restoration 3. Agronomy I. Title.

www.PRINTABLEREALITY.com

To my boy Bryn

&

my Nana Blanche

& to all the lovers, collaborators,
innovators, pollinators and inoculators
dedicated to healthy food systems.

FOR THE LOVE OF SOIL

Contents

Foreword

Dr Gwen Grelet

I became a soil molecular ecologist to better understand nature's wonders. I was also driven by the desire to share that knowledge, so we could all operate as competent stewards of our planet. However, by 2016, I had grown extremely uncomfortable in my career direction. I had lost my passion, joy and sense of purpose. At a time when it was so crucially needed, the science system was pulling me away from this sense of stewardship. Its reductionist approach increasingly appeared to me to lack in relevance to the real world. Its strategy of risk avoidance binds scientists to remain within their comfort zone, where beautiful discoveries rarely lie.

As I set out to change my research focus, I discovered regenerative agriculture -- a movement that appeared to fully embrace the complexity and life force of nature, working to enhance synergies between all living organisms, including humans. I was so excited to discover farming principles that considered the whole. It was in stark contrast to my previous work that systematically tried to isolate complex relationships into seemingly more manageable "units."

As I learned more, I identified several regenerative practitioners at the forefront of this innovation. I knew instinctively that for any regenerative agriculture research to deliver insightful and relevant knowledge, I needed to partner with these practitioners. That was what led me to Nicole Masters.

Nicole was a shining light in the regenerative agriculture movement in New Zealand and there was no way I could proceed without having her on board!

After trying for months to get hold of Nicole, I eventually signed up for one of her workshops, determined to (1) learn more about her approach to farming and (2) meet her in person. Little did I know that my internal mental model of soil ecology would be so thoroughly dismantled and reassembled!

How could one person touch down on so many of the unknown of the so-called soil "black box" (i.e. the myriads of organisms living below-ground, unbeknownst, unseen and often ignored by us fellow above-ground human creatures) with so much confidence and, dare I say, almost a pinch of magic. How could she?

I spent most of the time trawling scientific databases to verify what this woman was saying and whether her stories were indeed backed up by data and trustworthy scientific studies –- each time, the facts backed her claims. Indeed, this woman knew what she was talking about! In a way, the cohesive whole picture that Nicole so beautifully articulated, was extremely challenging for my academic mind, trained to only discuss observed facts. Yet I found it so refreshing and inspiring!

Later, we sat together to discuss the ways we approach our search for knowledge and our work and the importance of not only acknowledging. We also explored how we can embrace vulnerability and joy in what we do and this too, was such a refreshing outlook on life.

As Nicole writes this book, we are also working together on a number of projects. Projects which focus not only on the science of regenerative agriculture, but also on mindsets, people's culture and relationships with their community and the land.

This book is a wonderful compilation of her life's work and an amazing personification of Nicole's gifts - connecting soil health to human and environmental wellbeing in a weave that is broad, challenging and scientifically grounded. It manages to draw an inspiring picture of the rising number of people, communities and initiatives who are right now empowering themselves to regenerate our connectiveness with the land and a true sense of stewardship for our planet.

This book is a journey into the future of agriculture and food production. Enjoy the ride.

Introduction

I'm not normal.

The Oxford dictionary definition for normal reads: "Conforming to a standard; usual, average, common, typical, or expected," with synonyms that include ordinary, conventional, set and fixed. Yup, I'm good with not being normal. Growing up, I did not have a "normal" childhood or early 20's, and I'm grateful for a life far from the drudgeries of any 9 to 5 job. I flounder in pleasant company with pleasant conversations, sharing stories about what the neighbors are up to. What lights up my curiosity, is finding out about what really inspires you and the stories you share. I probably won't remember your children's names, sorry. Give me your soil tests, however and I'll remember those numbers for years to come.

The farmers, growers and ranchers I work alongside "ain't normal" either. They are open-minded, high performance, eager to develop their own skills and take on a challenge; and have no concerns about working alongside and taking advice from, educated, driven and opinionated women. If the market is great, they'll sell their whole herd, let the grass grow, earn some interest and then restock again a year later, when the market's down. These operators are innovators, experimenters, practical researchers, observers, alchemists and calculated risk takers. And they're blazing a trail to healthier soil, upon which our collective future rests.

The promises of the Green Revolution are now bearing their fruit; soil losses are escalating well beyond soil's ability to regenerate, with dramatic impacts on the environment, food nutrient density and upon human life. Scientists calculate that in the past 40 years, we have lost nearly a third of arable land to degradation and erosion and we may have as little as 60 harvests left before catastrophic collapse.[1] We are entering a perfect storm, with increasing reliance on fertilisers and pesticides, compounding climate stressors, reduced resilience in food production systems and poorer outcomes for food producers. This decline is happening when we need food security more than ever. However, instead of seeing these outcomes as a disaster, I see them as an opportunity to draw people's attention to what's not working and inspire action to turn this ship around!

Science offers great leaps forward in knowledge and its contribution is vital, however, the scientific method is imperfect when it comes to nature. Scientifically proving theories in complex confounded systems, like those found in nature, are neither cheap nor easy. Understanding whole systems is a challenge; if we had Bayer's budget, exciting outcomes for soil health, carbon drawdown and water quality issues, would all come about much more quickly. Society is addicted to the quick fix and think big projects. It's relatively simple to develop and market new technical innovations: "O look, how shiny" or "See, this genetically engineered organism glows in the dark," or "Here, enjoy this herbicide. You can drink it."

Researching and funding good management practices, does not attract the imagination of companies looking to create marketable products. What fascinates me and gives me assurance that a shift is coming, has been witnessing the breakthroughs from the people on the ground. It's not the "silver bullet" products, scientific research, or the latest technological gadgetry leading the way, instead it is the farmers, ranchers and growers, who are transforming soil health, using practical applications that work in the real world. If we wait for science to provide the answers, we will come up short.

I'm an agro-ecologist, educator and systems thinker with twenty years of experience working throughout Australasia and North America in what may be termed "regenerative agriculture," "biological agriculture" or "holistic farming." This book is not about labels. It is about the approaches, tools and thinking required to build soil and ecosystem health, food quality and profitability. You can call it what you want. I call it common sense.

In the 1970s, the label "sustainable" became synonymous with the environmental movement; those concerned about atomic testing and burgeoning waste, pollution and ecosystem destruction. Sustain-able can be literally translated to mean "maintain the same," which is no lofty aspiration! The modern use of the term sustainable, has become 'green-washed,' manipulated by chemical companies so the word now represents business as usual. Any goals which set to sustain degenerating or extractive systems, fall short on what is critically required today.

Regenerative agriculture proponents are being asked to define what the term "regenerative agriculture" means and how it can be measured and ultimately certified. Pinning down a broad process, with principles that aim to improve current ecosystem function, has both pros and cons. A clear definition is helpful for scientists, funding agencies, consumers and the marketers, to help to deepen understanding and to promote the approach. For me, regenerative agriculture calls forth terms around integrity, restoring natural cycles and transparency of food production systems. The term 'regenerate' encompasses ideas such as: renew, restore, grow, to bring new life to. Any definition of regenerative, therefore, requires that we know where our starting points are. Are you rebuilding, restoring and bringing more vigorous life to your land every season? This also means that your goals for measuring success are evolving.

This drive to define and label regenerative agriculture, reduces broad principles and practices to a set of boxes, and boxes concern me as they put a lid on innovation. Personally, I'm interested in how we keep expanding and transforming what's possible for landscapes and people in this evolving world?

I've been pondering too, how to communicate a term which encompasses the groups of people who grow food and fiber. The usual terms include: producers, operators, managers, enterprises and animal and plant husbandry. These are all 'doing' words, words of control, finances, machinery and masculinity. Don't get me wrong, I love the masculine, and I love the feminine, and both are required in food systems. We need a new way, however, to communicate about agriculture, which is not focused on the control, domination and ownership of resources. It is this thinking that has lead to the modern agricultural crisis. The new agriculturalists are adaptable, flexible and synergistic,[1] those who view themselves as active caretakers, custodians, stewards or collaborators. These terms are all ways of 'Being', and it is these ways of being, which engender actions on the ground.

Giving labels and terms does help to summarise and quickly communicate complex ideas. For ease, in this book, I'm going to use the term 'regenerative' and 'regenerators' to describe the mindset and actions of the farmers, ranchers and growers, who are bringing vibrant life back to their land.

Nearly all who live and work on the land, see themselves as stewards. Most wish to leave a legacy of value for their life efforts. There can be a crushing sense of disappointment, when this legacy comes up short: with increasing inputs, pests, weeds, degenerating water quality and the resulting decline in profitability. I see potential in these unfulfilled dreams, the gap between what someone is committed to and their actions on the ground. For me,

[1] A synergistic effect means the whole is greater than the sum of its parts.

working with people on the land, every word and visual clue, is part of a journey that has no final destination.

In this book, I share what inspired and enabled these innovative and intrepid regenerators, to step outside expected rural behaviours and the rewards their courage has reaped. These food producers are taking actions to imitate natural systems more closely. Through doing this, they are creating food production systems with reduced reliance on external inputs. With a focus on mimicking natural systems, they are rewarded with more efficient nutrient, carbon and water cycles, improved plant and animal health, nutrient density, reduced stress and ultimately, profitability. I'll also share minefields to avoid.

I'm often asked to write a soil book "for dummies," sorry you won't find it here. There's nothing dumb about farming, ranching or soil. In fact, those who work with land are some of the smartest, most capable, multi-talented bunch you'll find. Where else would you find a mechanic, a vet, an accountant, a builder, a butcher, a cook, a marketer all wrapped up into one? Which leaves little time for reading, I understand, but stick with me folks, for within these pages there lies gold, for you and for the health of our ecosystems.

In 1996, I went to university to study an Ecology degree and become a Great White shark researcher, instead I 'discovered' soil science. What enthralled me the most, was not the scientific principles, but wondering, why weren't more people soil geeks?! If you're interested in water quality, food security, fishery beds, climate change, food and human health, you really need to get interested in soil health. At the root of all of these large, complex issues, also known as "wicked problems," is our relationship with soil. Now for some, words like 'climate change' may cause your brain to fizz, my advice? Take what you need and discard what you don't.

There is a whole world under our feet, begging to be explored. It's a world that has been largely ignored and kept out of sight, until more recently. We're spending trillions, figuring out how to get to Mars, or discovering life on other planets, when we are barely comprehending the life we have on earth. There are huge financial investments looking at ways to communicate with extraterrestrial life, when we are yet to crack the language of dolphins and chimps, let alone soil microbes! Soil doesn't offer the same kind of sexy powerful feeling that blasting a big rocket into space does (well for most people anyway) ... I understand; it's difficult for most people to get excited about something invisible to the naked eye. Soil microbes are small and it's hard to imagine the dramatic impact the unseen can have on the entire planet. Until you see your own soil through a microscope, soil microbes seem like fairy dust and a leap of faith.

Modern agricultural practices, which arose during the Green Revolution, yielded great leaps in production; production which ultimately 'robbed Peter to pay Paul.' The costs to the environment and loss of soil and soil carbon have been long unaccounted for, and the debt collectors are now knocking on the door. The many 'solutions' offered by chemical suppliers, to address these consequences, are often reactions to symptoms. These challenges require proactive thinking and actions. In these pages, we will investigate how addressing issues at their root cause, supports an environment for health; effectively reducing the need for soluble salt fertilisers, herbicides, pesticides or fungicides. I'll also show how chemical inputs place producers on a treadmill, requiring increasing inputs, with decreasing resilience and profitability.

Those who grow food are experiencing increasing global pressures, as they battle against increasing costs, unfavourable markets and climatic variabilities. Now more than ever, agriculture needs to consider whole systems accounting and put the power and profits back into producers' hands. There is a perception in rural communities, that economics and the environment are uneasy bedfellows. These concerns are unfounded. The benefits from working in sync with nature are being demonstrated by the regenerators, the characters in this book. These producers represent a tiny snapshot of a global and rapidly growing movement.

This book was born out of a desire to communicate and share my coaching processes, what I look for and where I triage my actions. And it is by no means the only way to approach soil health. I aim to demystify a process which involves identifying the enabling factors to production and performance and uncovering points of leverage. Just as nature doesn't use a hammer to correct imbalances, we'll look at how 'tickling the system;' using as little pressure as possible, yields far greater soil and plant responses. In these pages, you'll read how possible and rewarding the journey to soil health, resilience and growing nutritious food can be and how you too, can regenerate your soil.

People are what put the "culture" into agriculture. This book includes stories from producers, many who grow food in the most extreme and picturesque environments in the world, and what led them to shift their thinking. Some of these learnings and soils observations may be from ranchers, growers, croppers and farmers, who will be in different eco-types than you, growing different crops or animals. I invite you to put any concerns such as, "it's easier for them in their landscapes, or this doesn't apply to me," all to one side. I have taken Montanan ranchers to visit croplands in Western Australia, croppers to visit compost facilities, or viticulturalists to see cover crops on sheep stations. What these enterprises all have in common is soil, and you'd be surprised how a different perspective may just provide a valuable snippet of learning.

The information contained in this book is not new thinking; people have been working with and writing about, regenerative practices for centuries. It is insightful to read the works of authors and farmers who held strong beliefs that we are part of nature, who understood that observation is key and saw how microbes and humus are the glue which holds society together. At the turn of the 20th Century, many authors, such as Louis Bromfield, Newman Turner, Rudolf Steiner, Lady Eve Balfour, Aldo Leopold and others, were raising their concerns about the industrialisation of food; concerns which are as valid today as they were 100 years ago.

As the eloquent Aldo Leopold once wrote, "We abuse land because we see it as a commodity belonging to us. When we see land as a community to which we belong, we may begin to use it with love and respect." The pioneering farmers these authors wrote about were producing higher yields and better-quality foods, while building soil health. Facts which have been lost to the sands of time. I am not proposing we head back in time, what we have today, is a richer understanding of the science behind what these innovators were discovering in the field.

I'm passionate about methods which help to reduce chemical use, but I'm not a fundamentalist. My focus is on getting to the root of why chemicals are required on any given enterprise. At my company, Integrity Soils, we work with producers who have been uneconomically propping up their plants and animals for decades. Through addressing root causes, we can save costly, and often unnecessary, chemical bills, becoming part of the solution to issues, which range well beyond the farm gate.

In 2009, I drew together leading scientists and regenerators, for New Zealand's first Soil Carbon Conference, to discuss and share research and real-world experiences from a "soils first" approach. We came under sustained attack in the rural papers, with editors and agronomists arguing that there was "no sound science" behind a focus on building soil health and profitably reducing inputs. Interestingly, the terms 'sound science' or 'junk science' were first coined by the tobacco giants to discredit the evidence that second-hand smoke causes disease. It later became a popular argument in support of hormones in beef and silicone gel breast implants. Researchers warned health professionals, that the "sound science" movement was not about science at all, but a very clever political discourse, with the aim to manipulate scientific standards to better "serve the corporate interests of their clients."[2]

I wish I'd been aware of the history of "sound science" in 2009. I would have handled our media messaging more assertively, with fewer sleepless nights over the injustice of it all. The "sound science" argument has left a mark on the New Zealand conventional farming community, which demands peer-reviewed multi-year, multi-paddock research, in its own region and climate, before people will even engage in discussion. Even then, it's usually still not enough to get them to look outside the box.

In rural communities, conformity engenders a sense of stability, trust and reliability in communities, where trusting neighbours can come down to life or death decisions. A willingness to step outside the norms, takes great courage and commitment. This desire to enforce normality and stability keeps our lives small, restricts innovation and ecological problem solving and in turn, is inadvertently destroying the soil upon which life on earth depends.

Originally, I was advised to write this book as a technical manual. I have lots of valuable reference books, they sit on shelves and mostly gather dust. What I hope to share with you instead, is the cultural world of regenerative ag, and put the technical aspects into something that is hopefully more accessible and real. In doing so, I have added more of my personal voice than I initially intended, to bring you along on my journey.

Before we begin, I'd like to share a pivotal aspect of my regenerative story. In 2000, my father, Brian, retired from his international world of long-haul flying, to follow his childhood dream of becoming a farmer. He brought a property in Katikati, New Zealand, at the foothills of the wild Kaimai Ranges, and I joined him, with a brand-new baby on my hip. I hope I was of some help to him, as we planned and planted 700 avocado trees and an orchard. We also took on an extensive wetland revegetation project with his new wife Gaye and began a small herd of red Gelbvieh (pronounced Gel-fee) cattle. Gelbvieh are considered triple-purpose cattle, for milk, meat and draught. For us, it was a vertical learning experience around the impacts of heavy animals on steep land and mothering genetics, that were dangerously good. Dad would come in at night, with his stories of survival, "I got between the cow and her calf and she pushed me forty feet along the ground. Lucky there was a fence right there and she pushed me through!" And then he'd laugh. There was more than one occasion, where the tractor ended up in a ditch; its brakes would fail regularly "don't worry, just drop the forks if you need to stop," he'd advise me. Life on a farm with greenhorns is never dull. The farm offered an amazing internship, and an idyllic place to raise my son.

Dad chose one of the best growing climates in the world. Our view included a fresh swimmable river by the house, haloed by dense olive-green native bush, as a counterbalance to the verdant grass. It's not easy finding work as a single parent in rural communities and the stay-at-home mothering role did not come naturally to me. One fortuitous day, my father stumbled across an advert titled: "Deceased worm farm estate." Curiosity sparked, we made the decision to buy the business, which catapulted my thinking and my life's focus. I began teaching worm farming in schools and speaking with growers and gardeners about mycorrhizae and carbon.

I'd found my ikigai.[2]

So many regenerators I meet, exemplify this ikigai state. With the worm farm business, I had a passion and a mission, however the circle "what you can be paid for" was a little lacking. I knew soil health was important, New Zealand just wasn't quite ready for the message! Somedays it feels like it's been a long push and pull uphill, to see soil health receive the attention it deserves. However, that worm is turning, and turning fast.

---

[2] *The Japanese word "Ikigai" is loosely translated as; "finding the purpose of your life," what is it that gets you springing out of bed in the morning? This sweet spot arises when all these different life aspects come together in balance.*

Ikigai graphic representation by Emmy van Deurzen.[3]

I've come a long distance, both in miles and learning, from the jade bush and misty damp mornings, directing worms in Dads derelict milking shed. These days, for now anyway, I travel from semi-arid ranches to workshops with my cattle horse as a companion. Based in an old 26-foot trailer, we've been traipsing across the North American West, for the past few years. As I travel, it's becoming clear, that the soil bug is rapidly catching on, as farmers, ranchers and scientists begin to share their remarkable stories. Regenerative agriculture is still but a small fraction of our food production, however, once you are inoculated with the commonsense benefits of soil health, there is no going back. It's time for soil health to go viral!

1

Leaving a Mark

"Fear or stupidity has always been the basis of most human actions."

- Albert Einstein

I was 10 years old when my father, who had been a pilot in the Royal New Zealand Airforce, left to join Cathay Pacific Airways in Hong Kong; this also meant Mum, my two brothers and I were moving to the city. We had been living on a small farmlet near Whenuapai, just north of Auckland, New Zealand, with a few goats, pigs, a horse and cats. My only understanding about the place was limited to the name printed on my plastic horse collection which read, "Made in Hong Kong." In 1985, it was the most densely populated place on the planet, with 6,300 people living in every square kilometer. Saying "culture shock" was an understatement.

As a young kiwi, I loved to walk around barefoot, proud of my toughened soles. Older locals lectured me about the dangers of "stuff on the ground." However, I was invincible and continued barefoot through the community for five years, before I was struck down by a mystery illness. A fever came on and my neck and shoulder muscles tensed up, I had a sore stomach and my heart was racing. When a rash broke out, I was rushed into hospital with suspected meningitis. A lumbar puncture revealed that I was in the clear. As the fever abated, I was discharged.

I may have been free to go, but for the next 15 years, I wasn't free of the near constant headaches, migraines, heat rashes, foggy brain and a total lack of interest in schooling, family or sports. These symptoms could all be attributed to the array of normal hormonal and somewhat loathsome teenage behaviors. I visited every type of body worker imaginable: Chiropractors, Osteo's, Acupuncturists, Ortho-Bionomists, Cranio-Sacral, Bowen and Massage Therapists. Some offered temporary relief, some offered nothing; one thing they all relieved me of was my bank balance.

I was told my C1-C2 vertebrate were fused and that there was little that could be done, until I met an orchardist in Hawkes Bay, New Zealand. He shared a breakthrough he'd had in managing chronic fatigue, by working with a chemical detox specialist in Auckland. At this point, I had nearly given up. I was 30 years old and felt I'd explored every avenue. This, I promised myself, would be my last foray into finding a solution.

I made the 6-hour drive up to Auckland, migraine ever-present, to meet with Dr Matt Tizard. The doctor was well into his 80s, yet his bright brilliant blue eyes led to me to trust that he knew a thing or two about health. Entering his homely clinic, one wall held floor to ceiling books, the other contained vials of labeled liquids, against the third wall, a radionics machine sat on a polished wooden table. This uses 'frequency' in its diagnosis, similar to dowsing (or water witching); all of which required a leap of faith for my analytical brain. He had procedures to test for multiple chemical, environmental and disease agents.

His radionic prognosis? Paraquat poisoning, possibly picked up through a cut on my foot while walking along a recently sprayed ditch in Hong Kong. Those elderly Cantonese women knew what they were on about after all. I had no idea what Paraquat was at that time. I had been working in the organic sector, in horticulture, viticulture and beef, and it had never come up in conversation.

Popular in Hong Kong, paraquat is a quick acting restricted herbicide, often used as an alternative to glyphosate (Round-up). It is a known eco-toxin, particularly in waterways and has been strongly linked to Parkinson's. This is why it has been banned in 38 countries, including China and in the EU since 2007. Interestingly, 2007 was the same year New Zealand lifted its ban. Paraquat enters the body through ingestion or open wounds, like cuts on feet. When inhaled, it can cause poisoning and lead to lung damage. In the early 1980s, U.S agencies sprayed paraquat to kill marijuana plants, leading to controversy about its potential impacts on pot smokers. In third world countries, it is a leading method for suicide. It is cheap, readily available, has no antidote and you only need one to two teaspoons to cause a nasty, torturous death.

Dr. Tizard was adamant the Paraquat had accumulated in my spine. There is conflicting science about Paraquat's ability to pass the blood-brain barrier. However, as an additional invitation to Paraquat, the lumbar puncture offered easy access into my spinal fluid where it wreaked havoc upon my life for the next fifteen years.

Dr. Tizard's treatment involved three sessions in a hyperbaric chamber with intravenous vitamin C, followed by a round of homeopathic support.[3] In the wake of the first treatment, I felt like I'd been hit by a truck. For the next ten days, a nasty, toxic-smelling, grey stuff dripped out my nose. Within days I had full neck mobility and my painfully tight muscles relaxed, leaving my Osteo baffled. What practitioners for decades had told me was a structural spinal problem, was in fact, chemical loading.

There are currently 171,476 words in use in the Oxford English dictionary. In comparison, there are over 20 million names for chemicals. Every 3 seconds, a new chemical is invented. Who knows what impact they'll have on the environment or on people? American Western agriculture, which holds itself as an example of maximizing yields, no longer innovates in the field, it's moved its field into the laboratory.

One concern I have about this chemical experiment we are conducting upon the planet, is that its effects are not acute. We humans are blindingly slow to respond to events which lag in their effects or happen through generations; what we ecologists and systems thinkers call "feedback loops." Several farmers and chemical applicators have shown surprise by how badly paraquat affected my health. "I used to spray it all over myself with no gear on, and I'm just fine," they'd say. And so, I ask them about their children and the tone changes. "Well, our daughter died from leukemia," or they mention autism, learning disorders and other chronic health issues. There is growing evidence that we are altering our genome through chemical exposures and, as epigenetics shows, this in turn, alters our children's and our children's children's genetic makeup. In a drive to mechanically understand what makes up the world, scientists focused on DNA. It's DNA which make up our genes; however, it's the epigenetic proteins which switch on gene expression. These proteins are directed by environmental signals, such as

[3] In the practice of administering intravenous vitamin C, Dr Tizard was struck off the medical register. He was just a little before his time. Finally, this treatment is receiving the recognition it deserves, being used again for cancer treatments, lyme disease, shingles and chronic fatigue. He was a visionary who saved many people's lives, at great personal cost.

diet, stress, lifestyle and toxins. Imagine this process like building a skyscraper, DNA gives you the blueprint, while the genes provide the building materials, it's epigenetics who are the foreman and construction team who put the building together. If something was to go awry, perhaps the construction team drank too much the night before, or the foreman cracked under pressure, then the building may not be structurally sound.

Research into the impact upon human health, from the background of low-level chemical exposure, is mounting. And it's damning. Take BPA, and its companion phthalate, used in plastic storage containers since the 1960s, linked to heart disease, endocrine disruption and diabetes in humans. In mice, early exposure to BPA changed the epigenetic markers related to obesity, metabolism and aging.[1] When inadvertently consumed by women pregnant with male fetuses, say from drinking out of a plastic bottle, these chemicals are major contributors to the dramatic reduction in male sperm counts that have occurred over the past 40 years.[2] Since 1980, each generation of males born, grows into maturity with half the sperm counts of their fathers. Perhaps we are inadvertently solving our own population crisis?

Unless you were raised in a log cabin and clothed in animal skins, you have been exposed to a myriad of the genome-altering chemicals. Even isolation offers no guarantee; testing of Inuit populations in the Polar North, came back with results laden with heavy metals and "persistent organic pollutants" such as DDT, PCB, toxaphene, chlordane and more. Despite the best of intentions and scientific know-how, the cost of "production at any cost," is now being paid for by all facets of society. Today, it is not the 19th century infectious disease we succumb to, instead it is the slow and insidious auto-immune disorders, cancers and chronic organ failures.

Listening to nature's clues and signals is a crucial skill in regenerating landscapes. Listening requires we use all our senses, including common sense, in relation to the choices we make on the land, and make for our own bodies. What I find the most interesting, from the impact of paraquat on my life, is that I had already embarked on a path dedicated to reducing chemicals in the environment; unaware I was carrying my own chemical burden. My body and heart knew. Intuition and our body's knowing, is incredibly undervalued in our current society. We poo-poo it like it's weird, or you're a hippie, or weak. We're not stronger when we ignore our intuition. Just the opposite. We become duller. We dull our senses and thereby limit our opportunities for a rich and insightful life that results from paying attention and syncing ourselves with nature.

The regenerators I meet listen to their gut. They constantly question their own assumptions and question management decisions: "Is paraquat, atrazine or glyphosate a food ingredient?" "Does this input improve the quality of my soil or crop, or does it degrade it?" "Does this practice make my stomach turn or my heart ache?" The time to shift from the chemical paradigm to nature's paradigm is well over-due. There is no doubt we humans are a resourceful bunch. It just makes sense to align with a partner who is much bigger, smarter, more efficient and far more creative at solving problems than we are.

So many answers lie in the soil. These vital and functioning soils bestow benefits, not just to the people on the land, but across all of society and the health of the planet.

2

The Unintended Consequences

"Everywhere we look, complex magic of nature blazes before our eyes."
Vincent van Gogh.

Boxing Day, 2004 saw the planet jolted deep to the heart. A magnitude 9.1 earthquake gashed open over 600 miles of sea floor near the Indonesian coast. Spawning a massive series of tsunamis up to 30m (100ft) high, which slammed into the coastlines of 11 countries, leaving ecosystems, transport systems and communities devastated. No technological warning systems were triggered and over 227,000 people lost their lives. After the waters receded and recovery began, an interesting phenomenon was observed: the indigenous people, whom many were concerned may have been wiped out, survived. In fact, the marine dwelling Moken people or 'sea gypsies,' suffered only one fatality.

The Moken live in the Andaman Sea, near the epicentre of the quake, and are considered to be one of the least "touched by modern civilization." Their children learn to swim before they can walk. Compared to you or me, their eyes have adapted to see twice as well underwater and they can even slow their heartrates to dive deep for long periods of time. The Moken have oral traditions which tell of a spirit of the sea, which "eats people." Before the tsunami hit, they became aware of unusual signs: from restless animals, dolphins heading out to sea, currents shifting, fish disappearing and the forest cicadas going silent. Moken on the coast, took their children and tourists to high ground. Those at sea, took their boats to deeper calmer waters. Burmese fishermen, who had been in the same area as the Moken, did not survive the waves. When asked why the Burmese did not listen to the signs, the Moken fishermen said, "they see only their squid, they looked at nothing. They don't know how to look."[1]

The Moken represent an older way of being in the land or sea, deeply connected to nature's rhythms. Their natural responses were more effective than the modern technological systems. Many of us have lost this intuitive connection to land. We face a choice today: we can continue to relax on the beach, or we can heed the warnings and respond pro-actively to the tsunami heading towards us. How we can adapt and respond to an uncertain future, is the biggest challenge facing humanity.

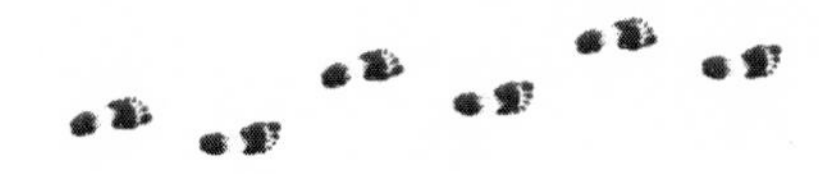

Most environmental degradation in evidence across the world, is not the result of intentional or malicious behavior, but the "unintended consequences." The best-intentioned of producers, often adopt new technology or quick fixes that carry hidden consequences. As humans, we've been adept at understanding complicated systems like machinery and computers. As ecological complexity thinkers, however, our current environment and state of human health globally, is testament to our failings.[2]

The crux of the issue is that food production systems are not complicated; they are complex. Many environmental impacts from agriculture are at a catchment or regional level and involve multiple stakeholders. These complex, or, in scientific terms, "wicked problems," have no simple fixes. Complex systems have interconnected parts, multiple factors involved and a change in one direction can unpredictably affect another. Complex problems include heavyweight issues like health care, gun laws, the obesity epidemic, "war on opioids," or climate change. Topics that will often get people worked up, yet no-one has the single solution. Opinions yes, solutions, not so much! Consider the complexity involved in raising a child; there is no one overarching factor in the outcome, and not one method that works with all children. Otherwise, the entire child-raising book industry would be out of business! By treating complex issues with complicated mechanical fixes, we end up with more problems. Mechanical thinking gives us solutions, like banning opioids or sugar, to address the systemic health issues or offering diet pills or methadone. These quick fixes do not deal with the root cause of why are you putting on weight? Is it hormonal, biological, epigenetics or emotional eating and how can we work to address those driving factors?

Nitrogen use in New Zealand (NZ), is one example of a 'wicked' complex problem which has crept up on society. Awareness around excess nitrogen (N) fertiliser, can be traced back to the early 1970s, when farmers and fishermen first noted algal blooms following aerial urea applications. In 1986, the NZ Government published a report, which included concerns around water quality and the need to reduce nitrogen use. No action was taken. Over the last decade, through the intensification of dairy and a 160% increase in N use, water quality in NZ has measurably dropped. New Zealand's "nitrogen balance increased more than in any other OECD country from 2000 to 2010."[3] I have been horrified when visiting poorly managed dairy farms, who "need" to apply over 300 units of N (per ha/yr), on lands located alongside braided river systems. Many lakes and rivers are now beyond a tipping point of "good health," due to nitrogen and phosphorus loading. In some areas, due to the slow percolation of nitrogen through deep aquifers, fertiliser applied in 1950s, is only just reaching rivers today. With increased awareness of the effects of nitrogen, efforts are underway to reduce N leaching from land. Most of the actions proposed, involve mechanical or technological solutions such as: genetic engineering, nitrogen fertiliser inhibitors, stand-off areas for livestock, precision irrigation and fencing off waterways.[4]

While these measures may help in the short term, applying technological fixes to biological issues, generally results in unintended consequences further downstream. For instance, the nitrogen inhibitors, used to stop the conversion of N into more volatile forms, washed into New Zealand streams to inhibit natural N-cycles. As a side-effect, ammonia was released, with far more negative impacts on water quality than nitrates. Then these chemical inhibitors were discovered in milk, and these products were immediately removed from the market. As with many issues that creep up slowly, the early warning signs were ignored. Today there has been little progress on mitigating the N issue. And the elephant in the room remains; why is N still being applied at such heavy rates?

How can we address many of the complex issues facing society and the planet today? We need to be smarter around "whole-systems accounting" of our resource use. Including all the costs from the start of an object or input; what is the full cost to extract a mineral or fossil fuel from the earth? Or the final cost to dispose of toxins or clean pollution up from the water or air? And even what are the costs on our medical health systems?

In nature, there is no such thing as 'waste.' When it comes to pollution, we modern humans have always believed in a magical place called "over there," a place which is out of sight and out of mind. First, we dumped into the streets, then to landfill, then deep underground, into the sea and even out into space. All these dumps are now under burgeoning pressure and our wasteful legacy of 'over there,' is finally catching up to us.

Farmers, gardeners and ranchers often ask me for a simple prescription to cure their ailment, be it a nitrogen issue, weeds or any other wicked problem. They want to be told: 'The Answer.' But just as there is no one way to raise a child, there is no single answer when thinking about soil. To find solutions to complex problems, new approaches and new thinking are required. What would nature do? Through a deepening of our relationship with nature's dynamics, we can enhance natural systems, reduce waste and reduce the need for external inputs.

Like my father, my grandfather on my mother's side, Captain Robert Allen, was a pilot. Also akin to my father, he was a handsome guy; a dead ringer to the famous actor Clark Gable, and I'm sure he caused quite a stir with the ladies in his day. We all knew the Captain affectionately as 'Grump.' He would let me steer the tractor in his post-flying days. I would pretend to be a farmer, while he would impart stories of his heyday as a top dresser, flying on fertiliser-in a Douglas DC-3 no less. This twin-engined aircraft revolutionised the air industry in the 1930s, carrying up to 32 passengers in comfort. As a bona fide hero, it carried troops into remote regions in WW2. The DC-3 was even used as a work horse in the Vietnam war; dropping propaganda to the masses and flares to illuminate the Viet Cong's hiding spots. I doubt that Donald Douglas ever pictured his beautiful aircraft being used to spread fertiliser. I often imagine Grump's days working for Fieldair in New Zealand, droning down through steep valleys and up across the golden ranges of the Hawkes Bay. It would've been an impressive sight.

In the 1960s, pilots were applying superphosphate, a phosphate rock treated with sulphuric acid. Phosphate is the 'P' in the triple chemical fertiliser triangle NPK. It was heavily subsidised by the New Zealand Government, to fast-track turning native bush areas into grass production. The source for this early phosphate rock, was high quality guano rock (ancient seabird poop), mined from the Pacific Islands, Nauru and Banaba, part of the Republic of Kiribati. Have you ever seen Banaba in any tourist brochure? It's doubtful you have. What was once an inhabited 6 km^2 island paradise is now a hollow shell.

From 1900-1980, phosphate rock mining stripped away 90% of the island's surface, a process mirrored on neighboring Nauru. The elevation of the landscape dropped 50-60 meters, before P reserves were exhausted. Following 80 years of partying, the hangover kicked in, and the miners moved along. The island now resembles a modern apocalypse, adorned with abandoned mining equipment, empty swimming pools, a rusty old movie projector and golf courses, relics of a frenzy of greed.

This exploitation paid for the rapid growth of New Zealand and Australia farming economies. The lesson, Dr Katerina Teaiwa, a Fiji-born academic, asks us to take from Banaba's history, is to consider what is behind the growth of commodities like beef, wheat, corn or dairy products. "This chain of fertiliser ...makes it possible for you to have mass agriculture," "I think it is important for humanity to think about what the trade-off is for all of these things we take for granted." Where and how was your food grown? Are these producers focused on growing nutrient dense foods? Or are they after cheap, low-quality commodities, where the costs are borne elsewhere?

Surely, we've learned from history? Apparently not. As easy phosphate sources become more difficult to extract, other regions opened up for mining in China, Florida, Morocco and Bou Craa in Western Sahara. Phosphate is an essential element in modern industrial farming systems and any risk to supply is a threat to global food security.[5] As such, phosphate is a bone of contention, and arguably one cause for Morocco to stamp its claim on Bou Craa and the Western Sahara. Add in potential oil reserves and valuable off-shore fisheries, and you have a trifecta to pique your neighbour's interests.[6] You've no-doubt heard of "blood diamonds;" similarly, warfare has broken out around P supplies, with the rock now referred to by some as "blood phosphate." The food you ate this week, probably contained this disputed rock.

Unfortunately, as the higher quality seams dry up, P from lower quality sources is now being shipped across the sea contaminated with elevated levels of heavy metals, cadmium (Cd), chromium (Cr), mercury (Hg) and lead (Pb) and radioactive elements like uranium (U). These elements accumulate in soil and are toxic to both humans and animals.[4]

Cadmium is linked to human diseases, including stomach ailments, osteoporosis, heart and kidney diseases, fertility problems and endometrial, breast and prostate cancers. Cadmium accumulates in major organs. In New Zealand, following the deaths of dogs in 1990 from eating contaminated meat, the Food Safety Authority placed restrictions on the supply and export of offal from animals over 30 months of age. "Fortuitously for the dairy industry," says freshwater scientist, Dr Mike Joy, "the cadmium ingested by cows doesn't get passed to milk, or we would have been banned from export markets decades ago." Dr Joy has taken on the role of a canary crying out in the mineshaft of New Zealand's declining water quality. He continues to take a stand for waterways, despite varied interests seeking to muzzle and discredit him as a scientist.

Every year, New Zealand farmers apply nearly two million tonnes of superphosphate, pouring 30 tonnes of cadmium onto the fertile volcanic farmland. Concerned about potential negative public reactions when land-use changes from agricultural to residential, the NZ government had the foresight to tackle the problem head on...by removing agricultural land from being officially declared as contaminated by cadmium. Genius. "Over there," is now under our feet. I wonder if the Cd had come through in the milk, perhaps New Zealand's environment would be in better shape?

Globally, we are using and abusing this finite resource at an alarming rate. Most of the P applied agriculturally, is destined to bind to or "lock-up" in soil, before moving with soil particles to contribute to the collapse of waterways and oceanic "dead zones" --vast swathes of ocean water devoid of life.

These dead zones are created through excess nutrient pollution, from human activities such as fertiliser and sewage, which feed algal blooms. In

---

[4] For reference, allowable Cd limits in the EU are 20ppm, Australia 131ppm, New Zealand 280ppm and in Canada, a whopping 889ppm.[4] Over half of the high Cd phosphate shipped from Morocco goes directly to Canada. This leaves me wondering -who gets to decide what level of Cd is acceptable for human, soil and animal health?

turn, these blooms deplete the sunlight and oxygen required to support most marine life, creating a mecca for jellyfish, and not much else. Jellyfish for supper anyone? These dead zones can be found on most inhabited coastlines around the world. At last measure, the dead zone in the Gulf of Mexico, which drains rivers from the US corn and wheat belt, set a record in 2017, when it covered more than 22,730 square kilometers (8,776 square miles), which is about the size of the state of New Jersey or larger than the country of Croatia[7]. With massive flood events across the U.S., the 2019 forecast is looking dire with fisheries asking lawmakers to declare a fisheries emergency. There are thousands of these dead zones around the world, covering nearly 260,000 square kilometers (100,000 sq. miles).[8] These areas are estimated to have resulted in the death or exodus of over 10 million tons of crustaceans, fish and mammals. Even 'clean green' New Zealand has a significant dead zone near Port Waikato.

Science is coming up empty on what to do about phosphate; looking at more and more mechanical technological fixes, for an issue nature solved about 240 million years ago.

While my Grandfather was artfully flying on superphosphate, native insects, including porina -- a giant voracious soil dwelling caterpillar --were having a literal field day on the introduced grass cultivars. In response to these insect pressures, the organochlorine DDT, was added into the superphosphate mix; why not kill two birds with one stone? DDT was great at killing a multitude of insects at once...and inadvertently wiping out thousands of fish and eels, and more than two birds. DDT accumulated right up the food chain, so you can still find it today in New Zealand orca, the most heavily contaminated marine mammals in the southern hemisphere. With a specialized diet of sediment-dwelling stingrays, orca are also full of fire retardants and PCB's (another banned substance that just won't go away.)

DDT's inventor was awarded a Nobel Prize, as it so effectively controlled malaria during WW2. By the late 50s, DDT drew Rachel Carson's, and the world's attention, in her breakthrough environmental book, Silent Spring. DDT was first banned in Hungary in 1968 and in the U.S. in 1972. And clean green New Zealand, the land of The Hobbit? It wasn't fully banned until 20 years after Hungary. Is there a theme starting to emerge here? The beautiful land I am proud to call home, appears as if it's trying to win a race to the bottom, while wearing a blindfold.

Now my Grump was a highly educated, empathetic and loving man who cared about people and the environment, yet he was inadvertently part of a sad chapter in New Zealand's history, which he regretted until the end of his days. And here we are, doing it all again; still not learning from the lessons of our elders and continuing to believe there's some mythical "over there." More unintended consequences for future generations to deal with. We don't need more intellect, knowledge, or computer power to address these impacts. To paraphrase Albert Einstein, it's the structural supports and thinking that has got us into the pickles we are facing in agriculture and it will take a shift in our thinking to discover long-term solutions.

Avoiding the 'unintended consequences,' starts with the use of all our senses. This includes tapping into our sense of wonder. We are now moving into the next phase of agriculture, what researchers are calling agro-ecological or "post-modern" agriculture.[9] This approach to food production powerfully addresses inefficiencies and waste through optimising inputs and management. It doesn't require more modern technology or expensive inputs. It does, however, require producers to hone their observations, thinking and learning skills to produce top-quality nutrient-dense foods. If we don't address the structures in agriculture and look only at changing practices, we'll continue to pass on these unintended consequences to future generations. To do this, requires a transformation in how we see, think and act in the food production spaces.

Let's be clear, transformation is not the same as change. Transformation does involve change, but not all change is transformative. When we think of change, we're often looking to make things better, improved, or different than they were in the past. These changes are external and can be easy, such as buying a new piece of equipment, growing a beard, or going on a diet. When you make a change, sometimes it's easy to go back; reach for that box of cookies or cut off your beard. Transformation on the other hand, is irreversible, a door that once opened, can never be closed again. The butterfly cannot return to the chrysalis. When a transformative shift happens, it's an internal process which shifts our core values and desires and usually involves an aha moment. The world appears different in some way and relationships shift. This might be a relationship to family, to yourself, or with your land.

Transformation can seem a little scary. That little voice in our heads, our ego, assures us that to stay the same and keep things familiar is somehow equated with safety. Anything unfamiliar and new can create a sense of anxiety, fear, apprehension, suspicion and avoidance. Your little voice was mistaken, thinking that the *unfamiliar* and *new* was risky; when in fact, the most risk lies from *not* looking outside the known. If you're not experiencing some sense of low-level anxiety, you're probably not living life to the full. It's in this slightly uncomfortable state where the good stuff lies. The only risk I see, is from those unwilling to take on regenerative practices. This book enquires into how producers powerfully navigate breakdowns and insights to transform their enterprises. This is a call for all of agriculture to heed the benefits of deepening our relationships to land, to be like the Moken people and not just focus on the squid before us.

FOR THE LOVE OF SOIL

3

To Bring Vibrant Life To

"Upon this handful of soil our survival depends. Husband it and it will grow our food, our fuel and our shelter and surround us with beauty. Abuse it and the soil will collapse and die, taking humanity with it." Sanskrit text ~1500BC

This is an exciting time for agro-ecologists and those who, like you, are passionate about soil. There are initiatives across the planet celebrating the intrinsic value of soil. In 2015, the "4 pour mille" initiative was raised by the French government just before COP 21 in Paris. It appeared that the worm was finally turning; soil and soil carbon were finally being recognized for their powerful role in drawing down carbon and other greenhouse gases. However, with scientists continuing to squabble over measurement techniques, global estimates and calculations, the necessary commitment to take global action stalled. It appears that putting initiatives into action, has largely been left to those on the ground.

In 2018, I was privileged to have Alannah McTiernan, the Western Australian (WA) Minister for Agriculture, open a soil workshop. Her stance on soil health was clear: "If Western Australia doesn't adopt regenerative agriculture, there will be no farming here in 10 years." A striking prophecy indeed. It's never been easy to farm in WA, with extreme temperatures, drought and frosts (not to mention the flies!). All projections point to even rougher farming ahead, without resilient healthy soil systems.

There's nothing like being a global traveler, for it to hit home how climatic extremes are now a reality across the world. We've all been witness to versions of the hottest, windiest, coldest, wettest and driest days on record.

Resilience is what gives us the ability to adapt or bounce back from these challenges, and this is critical now more than ever. Building soil health is the foundation to our ability to weather the oncoming storms. Having healthy soils, impacts on every facet of society and the environment.

So, what exactly is soil health? Most established agricultural labs define a healthy soil, as one which provides maximum yields; and they have decades of tests calibrated to yield to back this up. Sadly, this model of testing has proved inadequate for what producers, and society, require. Another definition considers soil health as the absence of pests or diseases, which is an impoverished view to hold indeed. Any definition of soil health must include profitability as well as crop quality. Definitions around soil health must consider, too, reductions in greenhouse gases emissions, buffers to toxins and resilience to climatic pressures. Healthy soils have functional water, carbon, decomposition and mineral cycles. These soils are vibrant and alive with full structural integrity, able to withstand heat, dry, floods and rebound quickly after disturbance.

The USDA defines a healthy soil as: "the continued capacity of soil to function as a vital living ecosystem that sustains plants, animals and humans." That this definition attributes the soil as living is a massive step forward in understanding. Particularly when you consider that most soil schools base their teachings on theories from mid-19th century scientist Justus von Liebig. Liebig is considered by many to be "the father of Chemical Agriculture." His theory was that all a crop required to grow was the big three elements: nitrogen, phosphate and potassium, aka NPK. He outright dismissed humus as having any role in plant growth. Liebig also put forth the "law on the minimum"; that a deficiency of any nutrient is the weak link in a chain, reducing uptake of other minerals. There are rumours that he came to regret his findings and later recanted; realising the intrinsic value of organic matter. However, by this time, a juggernaut of chemical companies were committed to ensuring their money train would not be derailed.

The significant scientific breakthroughs in the 20th century, transformed the way we observe the universe and ourselves. Revolutionary theories in quantum physics, mathematics, consciousness and the very nature of reality, deepened our scientific understanding to see that yes, everything is connected and yes, *everything* is energy. In 1942, the quantum scientist and mathematician Luigi Fantappie, coined the term "syntropy" to describe the attractive force which all natural systems gravitate towards: increasing order, complexity, structure and life. Regenerative soil systems are a demonstration of this principle in action. While many scientific fields took great leaps forward, the 'natural sciences' of agriculture, soil and agronomy continue to draw upon 19th century linear, reductionist models; seeing life as "machine." Through continuing to trust Liebig's doctrine, our entire soil and agricultural education systems are gravely flawed. Current microbial research speculates that most nutrient function, is mediated through and by, biology. Ignoring the basic premise that soils are alive, is costing society dearly.

A producer, well-versed in benchmarking and setting targets, is the energetic owner of Riversun Nurseries, Geoff Thorpe. Geoff has a youthful spirit, which belies his resolute drive for excellence. On first meeting you could be excused for thinking you've come across a surf bum, with his wild golden locks and sun crinkled smile. His passion for quality and pushing the boundaries of innovation, has not faded in his 35 years with Riversun.

The nursery is in Gisborne, on the eastern-most point of New Zealand's North Island; the first city in the world to see the rising sun. It is near here that James Cook first sighted New Zealand, and the Māori waka (canoe) Horouta first landed. Horouta brought the Kumara, or Sweet potato (*Lpomoe batatas*), which was cultivated and provided an essential and life-saving staple in the Māori peoples' diet. The warm, moist, temperate climate in Gisborne, gave life to the Kumara which had traveled for many months across the seas. It was these oceanic tides and ideal growing climate which kept Geoff, once an avid windsurfer, in the region too.

Riversun is New Zealand's premium supplier of grafted grape vines and rootstock to many of the country's top vineyards. In big production years, they have grafted around 5 million rootstock. They are also producing grafted avocado trees and kiwifruit vines in response to changing markets. Geoff has an eye on predicting market trends. In this business, divining the future can be the difference between success and bankruptcy; rootstock nurseries need to be able to predict what crops will be fashionable for planting the following season. In 2008, there was a major collapse in the grapevine nursery industry, which killed off a booming trade and thirty vine nurseries plummeted to just four.

In 2017, Geoff was made an Officer of The New Zealand Order of Merit (ONZM) acknowledging his services to the wine sector. His exuberance and passion for quality is contagious, a valuable trait with his team of 70 full-time staff who work across a diversity of enterprises. Enterprises which include a world-class laboratory and a team of scientists working on identifying diseases and running PLFA (Phospholipid Fatty Acid Analysis) test for soil health. The PLFA test measures fatty acids found in the cell membranes of living soil microbes. Many organisms produce specific or signature types of PLFA biomarkers, which can then provide an accurate fingerprint of soil microbes present and active in a soil.

For Geoff, quality and soil health are non-negotiable, number one drivers for his successful business. His shovel is a constant companion, he's dug more holes than most seasoned agronomists. We walk down the towering rows of the source blocks, under-planted in diverse soil building forbs, grasses and legumes. Digging under the vines, he takes soil aggregates, the large crumbs indicative of a healthy soil, and rolls them gently between his fingers. His dark brown soils have a colloidal stick, which leaves a stain on his hands. He draws the smell of soil deeply into his nostrils, before he knowingly smiles, "the secret is in the soil." This attention and love has provided the foundation for a hugely successful enterprise.

In the early 1980s, the nursery was operating largely along organic principles. However, the realities of providing certified virus and disease-free rootstock to a fastidious audience, meant the organic dream was shelved. The nurseries spent Winter months bare chemical fallowed. In Spring, their beautiful volcanic loams were cultivated into a fine tilth, laid with plastic, planted with slow release fertilisers, irrigated and then harvested at the end of the season. Plus, a soup of fungicides and organophosphate pesticides to keep diseases and vector insects, like mealy bug at bay. Over 30 years, Riversun's organic roots had eroded away, leaving them to chase their tail using chemical controls.

Seeing this decline, Geoff took stock in 2005 and brought together a crack soil and nursery team to ask the deeper questions; "how in a disturbed ecosystem could soil health be improved, build carbon and enhance a microbial community ideal for health vine growth?" Baselines were recorded around soil, finances, plus the in-depth data already collected in the rootstock blocks around growth patterns. Asking "why" questions is critical in leveraging regenerative decisions. Asking "why" at Riversun revealed a legacy of unexamined decisions. "If we can't keep doing this on the same land, in the same community and on the same planet for the next 100 years and the next 1,000 years, then it's not sustainable," says Geoff.

A common practice in nurseries is the placement of slow release fertiliser prills. Trialing alternative inputs, including blood and bone, chicken manure, fish and a control using no-inputs at all, the Riversun team soon discovered that they could discard the slow release prills without any reduction in plant health, vigour, or yield. The fact that the organic fertilisers cost less than the slow release fertilisers, was an added bonus.

After four years, of using only organic fertilisers, they have cut nitrogen use in half. What I enjoy about Riversun is their ability to trial new ideas with rigour; they have large scale trial blocks which have received NO added N for two years in a row now and have seen *zero* difference in vine colour, vigour, or yield. This is despite soil test results telling them there is no available N in the soil profile. Organic supplies of nitrogen are in fact the ultimate "slow release fertiliser."

Riversun has also been trialing several different green crops. These are grown every third year, as a rolling rest phase for the nursery soils. The goal is to eliminate the need for any N fertiliser additions to the soil, instead relying on leguminous crops and soil bacteria to harvest atmospheric N (after all, 78% of the air we breathe is nitrogen!)

The rationale behind the use of black plastic, was also questioned. The plastic is traditionally used in grapevine nurseries around the world to control weeds and to warm soils early in Spring. Unfortunately, it also stops water penetration, so irrigation tape is laid under every row, every year. As soon as the Summer sun arrives, this plastic becomes too hot, so labour was required to paint the black plastic white to prevent the very hot air under the plastic "ring-barking" the grafted cuttings. One block was trialed replacing plastic with barley straw. Rolling out small bales between the rows, saved the need for irrigation, plastic and paint. Soils could now breathe and even with the cost of labour and straw, the barley came out on top financially and environmentally. No need to dispose of the thousands of meters of irrigation tape or plastic either. The only downside in the first year, was that the barley seed sprouted, and a fungus was found growing up the trunks. As a result, the straw is now steam sterilized when it arrives.

Since embarking on a regenerative programme, focusing on vibrant soil health systems, Riversun has now eliminated herbicides from field nursery operations. With a response in plant health synthetic and systemic fungicides, all insecticides have also been removed from the nursery. As is often found with regenerative practices, ecology and economics, work hand in hand.

Update: the sheer logistics of growing, harvesting, pasteurizing and then spreading the straw (1,500 conventional bales/ha, 750/acre), not to mention trying to control late Autumn weed growth which appeared once the straw began to decompose, has seen Riversun focus its research effort into finding a genuinely biodegradable poly-mulch. They have found a specialist manufacturer in Europe and have trials underway in late 2019.

Many of the examples shared in this book are from degraded and challenging landscapes. How these soils can be regenerated is one of the most significant challenges for our generation. The producers in this book are regenerating their soils, restoring water cycles and bringing vitality back to ecosystem "patients" who have flatlined. Their success lies with the ability to identify critical points of leverage, a literal triage for addressing what can be termed the "limiting" or "enabling factors." When considering what the number one factor for growth is, most producers will think of water. Water is so immediate and in low rainfall times every producer sees the instant response from a rainfall event. However, that is not the first limit to production. The main driver for production is the sun, then air, then water. If the sun goes out, we are all in big trouble! How long can you survive without air, without water or without food? It is the same for our soil systems, microbiology and plants. If soils are tight and compacted, struggling to breathe, this is our starting point in the triage. Once these areas are addressed, then our triage looks to the next enabling factors: water, decomposition, then mineral availability. Too often, agricultural advisers look to the fourth step --the Liebig NPK. These advisers train their producers to reach for a bag, as the first step in growing crops.

We're going to cover each of these enabling factors through the next chapters. I'd like you to consider for now, that each of these factors are influenced by what we call the 5 M's: Mindset, Management, Microbes, Minerals and OM (Organic Matter)-- (yes, OK I'm cheating a little with the OM!). In each chapter, we'll look at how the 5 M's interact and interrelate to influence soil health and your goals. The actions you take around addressing these factors are governed by simple soil health principles. These principles are based on how soils work in nature.

Soil Health Principles

* Maintain soil groundcover and protection.
* Living roots for as long as possible.
* Incorporate livestock and/or their manures (where feasible).
* Diversity, diversity, diversity.
* Optimise plant photosynthesis.
* Reduce disturbance - minimise killing your underground livestock.
* Manage for what you want, not what you don't want.
* The actions which arise from these principles are influenced by your specific climate and circumstances.

Take your pick on what motivates regenerators; their individual drivers are as variable as the people on the land. Personal drivers range from increasing profitability and resilience to community or family concerns. Some are concerned about environmental or food quality declines. Others are looking for community recognition or to be the best producers possible.

Regenerative Agriculture provides benefits across a diverse spectrum to support producers in fulfilling their individual goals.

Of the 5 M's, mindset can be the biggest drag or driver for success. For example, if you believe that things are out of your control or that change is hard, guess what? "If you think you can or think you can't. You'll be right!" This attitude is what becomes a reflection of your reality on the land and in life. I find, even when initially working alongside clients highly reliant on chemical inputs, if they are curious and have an open and inquiring mind then their soil and crop programme is much more likely to be successful.

In our soil coaching work, we've found that one of the biggest determinants to success is a person's mindset about learning and failure. A 'fixed mindset' believes that intelligence and talent are what you are born with. Having this mindset, leads a person to look for external influences, to blame others, the weather, or time, when things don't go well. Any failures or criticisms are seen as failures of self, rather than a failure to produce a desired outcome. People with fixed mindsets, are often threatened by other people's success; this can often be seen in smaller rural communities, where doing something different or standing out as different is frowned upon. This mindset can be the biggest limitation to innovation and success. These are not the people we choose to work alongside. In contrast, people with growth mindsets, believe they can learn from experiences, that it's their attitude and efforts that determine outcomes. As the brilliant and sadly now deceased, dairy farmer Neil Armitage used to say about optimising soil health, "good, better, best, never let it rest." He was never content with being an average producer. His love for his cows and his land, meant he would never accept just a "good" outcome. Neil was always looking for the next step to harmonise and refine his grazing, grass species or inputs. As a result, in over 14 years running a low-input, high production system, a baffled chemical agronomist once rated his pasture "an 11 out of 10."[10]

There is a blurry grey line between a fixed or growth mindset. You may swing between a fixed or a growth mindset in some areas of life e.g. towards yourself, your land, politics or certain family members?! People with growth mindsets view challenges, including failures, as an opportunity for growth. I love hearing Gabe Brown, the North Dakota soil regenerator, commenting at farming conferences. "We plan to fail at least once a year," he says, "if we're not failing, we're not trying enough new things." It's not so much failure, as the willingness to try new things that produces extraordinary results.

A few years ago, we had an unusual interaction with a new cropping client; let's call him Farmer C. For over a decade, Farmer C had been trialing different inputs and cover crops and felt nothing was working. Every action we proposed met his resistance. "That wouldn't work in my soil." Or "I've tried that. Didn't work." He was clear. Nothing was ever going to change. In Client C's mind, the lack of success was no reflection on his actions, but due to external factors, like weather, seed choices, or other people. He was also concerned about any risk of failure; in his view failures would not be learning opportunities, but rather a personal failure of himself.

When prompted to share his cropping programme, it took some time before he sheepishly revealed; "Well, we apply a pre-emergent herbicide, then a broadleaf spray after germination, there's neonicotinoids on the seed, six fungicides" and lastly "atrazine to desiccate the crop at harvest." He then asked, "Can I sell this as biological?" I didn't have to ponder for too long to provide an answer on that! Farmer C had been adding commercial biological inoculants to his seed, including products for disease control, frost protection, water, phosphate and nitrogen uptake. These organisms all exist naturally in healthy soils. If soils are highly disturbed, they may be low in these critters; in those circumstances, addition of these products can provide a valuable opportunity to kick-start a programme.

The real issue for Farmer C? He was so concerned about letting go of the branch he felt safe on, that he couldn't take the actions required to swing to a new branch of actions. Through conversations, we discovered that he had never trialed anything for a 3-year period, in the same place. He had no baseline or methods to assess if soil and plant health was improving or if he was even achieving his goals. He also had no measures to assess if his management was regenerating or degenerating! Farmer C had not set himself up for success; his only measure was yield. When no immediate yield changes were seen, he would abandon a practice rather than experience a potential 'failure'. When he asked, "What do you think my limiting factor is?" I replied somewhat reluctantly; "Your attitude."

Over the next two years, Farmer C followed up on his trials and reduced his fungicides and pesticides. He saw the opportunity to produce cover crop seed for an expanding niche market. He began to see how his attitude had stopped him from following through or sticking with a new technique. He also saw how his concerns about his neighbours, and fear of risk meant he was unable to achieve his goals to reduce his chemical inputs.

Fixed mindsets make working with regenerative methods challenging, but not impossible. Despite the term "fixed," this way of seeing the world does not have to be permanent. You can change your mindset too! It has been inspirational to watch a community grow around him, that is hungry for the knowledge he has gained over these years. It's been an absolute joy to see the changes in him, as he has developed courage and confidence in taking actions.

One day, while visiting a client, I rode in the passenger seat of an old beat-up farm truck. I had the rare luck of never once having to open any gates; they were all lying on the ground and we drove right over the top of them! The farm resembled the typical tourist postcard for New Zealand, with green rolling hills, sheep dotted across every field and white capped mountains in the background. When he asked my coaching advice, it was: "close the gates and put in water!" There's no point throwing expensive inputs into a system, if you're not going to manage well. This also applies to any sectors, whether you're in cropping, home gardening or growing trees.

Management and timing are your number one tools, whether that's being adaptive with grazing, avoiding heavy traffic, not cultivating in the wet, using observations rather than calendar spraying, timing for pruning, planting seed and applying fertiliser. Without good management, you're putting the brakes on regenerating your soils. Unfortunately, this book does not have the scope to cover all the different management techniques for every sector. However, there are brilliant options available for further information and learning. For graziers, there are excellent books by authors such as Allan Savory, Jim Gerrish and Greg Judy, who cover the practical dynamics superbly well. If you've got livestock and you're not already adopting the principles advocated in adaptive planned grazing models, then that is the first place to start. For orchards and mixed enterprises, Bill Mollison, David Holmgren, Michael Phillips, Mark Shepherd, Gabe Brown and Joel Salatin cover everything from practical tips to the invaluable benefits of stacked enterprises.

I work with many zero-input adaptive management graziers, who believe that good grazing management alone, can solve everything. From what I've seen, the answer is not so clear cut --perhaps, maybe, possibly...it's a question of how much time have you got and what state is your resource in? Once we address enabling factors, grazing management is always your number one tool. In the livestock business, I'm seeing many long-term 'holistic' or 'adaptive' properties hitting walls. Initially, the change in grazing produced leaps forward in production, but then, over time, production stalls with some landscapes becoming weedier, more compacted and begin to lose animal performance. If they are monitoring well (which is one principle of Allan Savory's Holistic Management), then adaptive changes were required long before they reached this point. For landscapes which have plateaued, it's time to address what other enabling factors there may be, or indeed revisit and look more closely again at management, as something isn't working as well as it could.

Our management, or mismanagement, has had significant impact across the world. The arrival of humans into new lands, can be traced through archaeological records. We have been masters in modifying landscapes and biodiversity, wherever our feet took us; from early aborigines adopting fire and exterminating the megaherbivores, New Zealand Māori wiping out the gargantuan land bird, the Moa, or America's First Nation people playing a role in the extinction of rhinoceroses, camels, horses and even the American cheetah some 12,000 years ago. In more recent times, there have been massive historic shocks to grasslands all over the planet due to poor management, diseases, exotic invaders and common land grazing. We humans can't help but leave our fingerprints across a landscape.

What struck me when I first visited California, is the glaring lack of Summer groundcover. A look back into history books, is revealing. In 1880, prior to a devastating flood and subsequent drought, there were an estimated 500,000 elk, 2 million sheep and 1.5 million cattle in California grazing upon the lush and ample grasslands. During a 3-year period over a million animals died, bringing livestock businesses to their knees. Through the removal of every blade of grass during this drought, palatable native warm season perennial grasses and forbs dramatically declined. With the loss of habitat and an open season on hunting, so too did the elk, crashing numbers from half a million to under 3,500, one hundred years later. In Australia, another brittle environment, a similar grazing dynamic unfolded in a 2-3-year process, and now the green Summer cover historically available is all but gone.

Soils need cover and protection all year round. The best management in the world won't help if soils are exposed for nearly half the year. Introducing diverse deep-rooted Summer dry perennials, will protect and improve soils, open-up compaction and bring up minerals. I have no philosophical reservations about using native and non-native species. In my mind it is an absolute priority that these landscapes re-introduce Summer covers. What we are seeing around that world, is that as soil health processes begin to kick in, native species not seen for decades begin to appear again.

Sadly, many rangeland managers around North America and Australia are growing grass on the subsoil left behind by historic poor management decisions. Through poor management and overgrazing, topsoil blew or washed away rapidly. It is possible to restore function to these lands over time with grazing. If you're like me, however, you're just not that patient!

The high impact, longer recovery model proposed by many grazing advocates is challenging when the environments produce little ground cover. I see many graziers in semi-arid environments, still struggling after decades of adhering to this system. There are multiple challenges when you try to kick start the composting process. Many graziers have significant compaction or risk creating dust bowls when adding high concentrations of animals, without adequate groundcover and root systems.

One rancher who has used the adaptive grazing principles for over 30 years is Steve Charter. His "2 lazy 2" Ranch is located halfway between Billings and Roundup Montana. This land conjures pictures of the old West, which, before I arrived here in 2013, was country I'd only imagined existed in western novels. Stark rocky buttes interrupt the dusty open grasslands, which are dotted with stubby grey/green sagebrush and the occasional antelope. Billings was a town founded on cattle and its rich reserves of coal and oil. Steve runs 200-400 cow calf pairs on two separate blocks which cover over 8000 acres. The southern Winter block is drier with easier terrain, while the Summer block is a good 2 days trail ride to the nearby Bull Mountains with its stunning buttes, lookouts and protected valleys.

Steve has spent long days and weeks alone in these landscapes, with his cattle dog and horses for companionship. On my first visit to the Ranch, we tacked up in the dark and rode to the top of the home ranch. "The cows are somewhere in here," he gestured out across a thousand acres of steep buttes, screes, bull pine and juniper. "You go that way, and I'll meet you at the gate." And then he was gone. Luckily for me, or for Steve, finding and moving cattle fills my heart with delight. Five hours later, covered in dust, we met up again, with most of the mob, and my wits, collected. Riding with Steve has taught me many things, the lessons of which are evident in my saddlebag; a pocketknife, Chapstick, hoof pick, duct tape, sunscreen, snacks and water. My dad taught me, "We drink when the horses drink."" This may be fair; however, I don't have the constitution to drink what the horses are drinking... I've learnt to carry my own water.

Steve has never been one to accept the status-quo, driven by a deep need to do well by his land and to leave it in better shape. His parents were outspoken opponents of the coal industry, whose lands they border and lease. While Steve comes across as a laid-back soft-spoken man, growing up in a family fighting Montana's staple industry, he has ruffled some feathers. He turns animated when the topic of coal or ranchers being swindled by the very industry bodies established to support them. Neighbourly relations here, have been tenuous and litigious at times. Photos circulated online of Steve standing in a huge crack, remnants of the underground longwall mining process, which collapses tunnels behind the excavators, leaving the land above to recede and crack. These photos went viral, and the coal company directed their resentment towards Steve. They've taken him to court multiple times, sent sheriffs to his door demanding his personal emails and had their cases thrown out by the judge for harassment more than once.

The company wants him gone and continue to take hostile measures to push him out. Last season he arrived at his coalmine lease land and found coal company excavators deep at work, removing his dams and water pipes. His cattle had been without water for several days in conditions over 30°C (90°F). He's spent this season running to stand still with water pumps and generators to provide basic supplies to his cows. Yet he is resolute to keep his peaceful protest.

When Steve first took over the family ranch, the Soil Conservation Service provided the mainstay of information; advice Steve found to be "unsatisfactory." "They wanted us to plow native pastures and plant in crested wheat, which didn't sound right to me." Their advice to the ranching community? "They said, "Don't graze early," "Don't graze when it's going to seed, or when it's putting down root reserves." Which left Steve wondering... "That's practically the entire year. When exactly do you graze?! Soon after taking over the ranch, the area was hit by a prolonged drought. Steve muses now that if he had followed their advice, the ranch and its soil, would have all been blown away.

Luckily, his questioning of the status quo led to a chance encounter in the early '80s with Allan Savory, of Holistic Management fame. They became early adopters of the Savory grazing methods, with Steve's wife Jeanne establishing the first Holistic Management group in the US. This approach advocates higher animal impact and longer grazing recoveries. Before meeting Savory, cattle would remain in favoured areas, overgrazing the meadow floors and ignoring places further from water and on steep faces. Steve began to split up the large and open meadows, laboriously laying electric fences - many over a mile long to manage grass more effectively. Wild elk and deer offer a unique challenge to maintaining electric fences, keeping Steve occupied as they regularly leap, tangle, tear and drag fences down through the pastures. The high meadows responded to the change in grazing management with the emergence of diverse and dense plant cover, providing excellent Summer feed. After the snows melt, these high meadows get a good Spring grass flush and experience far less searing heat and drying winds than his lowlands Winter grounds.

His lowland pastures tell an entirely different story, however. The land had been cultivated and was one of the casualties in the era of the great 1930s dust bowl. The evidence of this period is still in evidence with deep mounds of soil along fencelines and gouged out dry erosion gullies. Calling soil 'dirt' here would be appropriate, as all that has been left is a structureless blow dirt, which travelled here from upwind neighbours. Over the past 30 years, with improved grazing management, this semi-arid grassland demonstrated only minimal change. On the brittle sagebrush lowlands, cattle are moved at least every 5 days to new pastures and won't return until there is adequate grass recovery - this rotation can be a minimum of 2 years. The land has improved, but most days Steve felt like he was taking two steps forward and two steps back. Close to the homestead, the land was dominated by moribund crested wheat and sage, surrounded by a sea of bare ground. He'd begun to wonder if the potential for the land had been met and "this potential," he said, "wasn't great."

Steve is curious about biomimicry and how to replicate nature's cycles more closely. He questioned, "What would the natural groundcover and grazing patterns once have been?" It's hard to find any conclusive evidence. Looking back into historic records there are some clues from buffalo jumps and First Nation stories. These clues point to possible bison migration pattern which may have only been every 5-7 years. Data for the migratory activity of bison is challenging to reconstruct. Historic use of buffalo jumps close by, show that these sites were not used every year by First Nation people. Large herds followed the good grass years, not returning every 3 months, or even the 2-year rotations that Steve's been practicing. This migration would have allowed for ample recovery and then high animal impact, in effect creating a layer of litter and compostable materials.

At the turn of the last century, poor grazing and the plough, led to the lowlands losing much of their species diversity, carbon and topsoil. The predominant short grasses, like blue grama, buffalo and the midgrass species, needle and thread, prairie junegrass, sand dropseed and western wheatgrass all but disappeared. Now, a deteriorating biological situation sets the scene, for "sleepy" desertified soils, perfect for sagebrush, cryptogams (non-seedbearing plants) and bare soil. Leaving these areas ungrazed for long periods of time, is no longer an option. Today, Steve's ranch does not experience the conditions or have the species that create flushes of tall grass that could be trampled. Not all management practices fit every ecosystem, particularly now that so many are so degraded. I've seen American cowboys turn New Zealand soils to muck and ruin. NZ farmers turn US lands to dust. Listen to your landscapes and research their past; there is never one prescription that fits every case.

For Steve it was his switch in mindset, from focusing on the aboveground production, to what was happening belowground, that unleashed the potential on his ranch. In 2014, Steve started to take the actions I only dreamed were possible on extensive rangeland. He started to spray biological stimulants and mineral catalysts to wake up soil life.

Steve built a custom 'slurry' sprayer and mounted it onto an old army truck. These slurry sprayers are designed to overcome the challenge of applying larger materials, such as compost, lime and seeds, in liquid suspension. In the first year, over 400 acres, he trialed 4 different treatments with mixes which included molasses, fish, seaweed, rock salt and vermicast (the fancy name for worm compost). For Steve, the results have been fantastic. In the first year, the quality of the crested wheat began to lift and by the second year, native grasses began to reappear. In this community, ranchers have been looking at ways to reduce the monoculture dominance of crested wheat. Steve is now seeing it naturally becoming out competed by Western wheat, the pasture sward is thickening, and the incidence of bare soil has been reduced.

Steve's cowhand Jodie, born and raised in this country, reflects that he'd seen big changes where the treatments had gone on. "I've never seen grass like this," he says, "in places it's knee high to waist height!" Jodie's caught the soil bug! Improving forage nutrients also means there is more grass in front of the cattle, enabling stock numbers to increase in the future, more grass to be trampled and more nutrient and water cycling underground. Steve cautions those wishing to adopt Holistic Management methods by doubling numbers for impact; in this environment that could be catastrophic; "You gotta have the grass first, then increase your numbers."

Steve has also been focusing on lifting plant diversity by adding seed to his cow's mineral supplement. When riding to check cattle, Steve sprinkles forage kochia seeds, leaving a trail behind him like a mounted Jonny Appleseed. Forage kochia (*Kochia prostrata*) is a small perennial shrub that has a definite place on arid and semi-arid rangelands. Interestingly, seeds struck better on poorer soils. The shrubs germinated superbly well around prairie dog towns, where they inadvertently planted the seeds around their burrows. The plants now grow taller than the prairie dogs can graze; making conditions slowly unsuitable for the wee rodents, who are now moving to the neighbours. Through action from the bio-stimulants, the once locally extinct grasses are returning in droves. It's an experience to ride out with Steve, as he knows his grasses. He'll leap from his horse in delight, pointing out emerging populations of native grasses. Grasses like needle and thread and western wheat, prolific before sheep arrived, are once again returning to his land.

Steve was so taken by the early results in the trial areas that in 2015 he set about building his own worm farm. Through the postal service, $500 worth of worms arrived and were bedded into 100 meters of windrows made from yard straw. The worms were fed a diverse diet of leaves, sugarbeet waste, manure, hay and wood chips. A perfect blend to grow beneficial bacteria, fungi, protists and nematodes. Over Winter, under deep snow, the worms kept working away, covered by a Geotextile cover and haybales. By the end of 2017, the worm farm was producing around 30 tonnes (66,000 pounds) of a quality biologically diverse worm elixir. Most was sold to progressive cropping operations who see the benefits from the microbial catalyst provided by vermicast extract. There was a plan to feed pumpkins too, from waste delivered after Halloween; instead it helped to feed a new generation of marauding deer.

Buoyed by the changes in his pastures, in 2017, Steve treated an additional 3,000 lowlands acres, as well as venturing into the rugged Bull mountains to cover another 1,000 acres in his custom army truck sprayer. After years of struggling to achieve the results that he had expected would come with changes in grazing practices, by addressing his lands enabling factor, microbial life, his management is now able to take his lowland pastures to the next level. "I now feel more empowered and positive about what is possible here and for all of agriculture," Steve says. He is clear that a paradigm shift is needed across society, if we wish to interrupt this cycle of moving from one disaster to the next. "Once we co-operate and work with nature, then things really start working." It's this mindset around learning and land regeneration, which enables Steve to take steps and experiment in a landscape which many see as too degraded and risky to afford to make the changes required.

4

Your Underground Livestock

"If you don't like bacteria, you're on the wrong planet."
— Stewart Brand, Editor for the Whole Earth Catalogue

We know nothing. Well, maybe we know a bit, like 1% of 1% of 1%. Our knowledge is likened to standing on the edge of a void, shining a flashlight into the dark. What our light illuminates, is only possible through our present-day understanding, technology and the questions we ask. What scientists are clear about, is we know far less about microbes, than what we do know. Current biological estimates[5] indicate that there are over 36 trillion species on the planet. Of these, around one million species have been mapped. You'll hear quotes like "we know 5% of what's happening in soil" when in fact it's a moving target, and with every new discovery, another doorway of knowledge cracks open. When most people argue for biodiversity, very few consider that most of the diversity is microbial and located under our feet. Soil is the most important and essential ecosystem, linked to every function on the planet. It is a frontier of discovery; from climate change, to keys to human health, nutrition and water cycling. Soils are an incredibly exciting field of discovery to be involved in.

There can be billions of organisms in a handful of soil, invisible to the naked eye. If you could picture yourself as a nematode winding through the soil environment, what meets you is a bustling metropolis, full of things to eat and predators to avoid. These city streets are not safe places to walk; they

---

5 Based on the current assumption that all biology comes from one common ancestor.

are dynamic and vibrant communities. If we had the equipment to listen into the soil ecosystem, it might sound like wild chatter and bloodcurdling screams! Many of the organisms in soil, make great subject material for a horror movie, like; Aliens and The Gift (Cordyceps mushrooms which burst from the bodies of insects), The Blob (slime molds), The day of the Triffids (sting nematodes) and Frankenstein (Tardigrade). Down in the soil, away from light, there is no need for eyes. Instead, microbes have incredible sensory systems; are you food, friend, or foe? Microbes stream chemical and electrical signals to communicate with each other, and with their vital support system, the plants.

There are very few "good" or "bad" microbes, despite them being framed this way since Louis Pasteur's pioneer work on "germ theory" in the 1850s. Disease organisms are present in every ecosystem and can often have positive benefits to plants and animals. E coli (Escherichia coli) for instance, often mentioned in panicked news bulletins recalling lettuce and spinach, offer immense benefits to us from birth, protecting us from other disease organisms and synthesising vitamin K and B12. Research facilities and pharmaceutical companies continue down a narrow avenue, focusing on pathogens and how to kill them. All this time, our questions have been misguided. The questions we need to ask are: what are the conditions that invite microbes to come and clean up? Diseases are often secondary to some other imbalance.

An elderly Australian stockman once told me; "You got livestock, you got deadstock." This rule holds as true for sheep, as it does to your microbial herd. If you want to keep your livestock alive and profitable, you need to follow the same basic requirements as sheep; they need air, water, food and shelter.

Consider the soil functions much like your gut: first you chew your food to make it smaller and then microbes and enzymes break it down further so nutrients can enter the bloodstream. The same action occurs in the soil, first by insects, ants, dung beetles and earthworms, and then by an array of increasingly smaller organisms, from nematodes to protists, fungi and then to the tiny bacteria. Just as a dose of antibiotics, stress or processed foods can give you digestive upsets, the same occurs underground following what we call "disturbance events." These may include natural events, such as flooding and fire, or human induced disturbances such as overgrazing, cultivation or herbicides, soluble salt fertilisers, fungicides and pesticides—what we call modern agriculture. Just as scientists are now discovering most of the health issues plaguing society are due to human gut issues, the same is happening in our soil systems. They have indigestion, gas, constipation and even rampant diarrhea! How we can rebuild the microbial bridge and optimise ecosystem health, is the biggest challenge facing society today. In a healthy functional soil, there are 6 major livestock classes: viruses, bacteria and archaea, fungi, protists, nematodes, micro-arthropods and micro-animals.

VIRUS

In this microbial grouping, I've included the viruses; these are sub-microscopic parasites, only able to replicate inside the bodies of others. Although they are not technically microbes, they make up the largest number of biological entities on the planet, so shouldn't be overlooked. In each teaspoon of soil, there may be over 10 billion viruses. Nature magazine states that: "If all the $1^{\times 1031}$ viruses on earth were laid end to end, they would stretch for 100 million light years." Collectively, all the viruses on earth weigh more than 1.6 billion elephants. The point being there are a lot of viruses. Yet we know very little about their roles in soil, they're "an unknown quantity within an unexplored territory." Like bacteria, fungi and insects, we know far more about pathogens, than we do about the beneficials.

Recent scientific breakthroughs now believe that viruses play a central role in the control of bacterial populations, the cycling of nutrients, carbon and the movement of energy in soil. Through horizontal gene transfer, they are the mediators for microbial evolution, and may have been the precursors for all life on the planet. Traditionally, the cycling of nitrogen, carbon and nutrients, was believed to be due to bacterial and fungal death, but research shows that this process did not fully capture the volume of cycling. If you were to remove predation by other organisms in the food web, viruses account for 100% of bacterial death rates and release of C and nutrients. These viruses maintain the youthful age of bacteria, in what scientist's coin as the "forever young" concept, ensuring a high turnover of bacterial populations. Research in China, comparing organic to conventionally fertilized fields, showed a 4-to-5-fold increase in virus populations in the organic fields. If you sift through the literature, all soil microbial conclusions end the same: "To understand the soil..., much work remains."

BACTERIA & ARCHAEA

Bacteria and archaea are prokaryotes, the smallest (<10um) and simplest single-celled organisms; Earth's original terra-formers. They are the first microbes to arrive onto the scene: onto the planet an eon ago, to any rock and leaf surfaces. Like viruses, they're found in every environment on the planet, from ice shelves, deserts, sulphuric hot springs, the air and in -- and on -- you. Both are single celled; however, archaea differ in their cell wall structure, and are oddly more closely related to plants and animals than bacteria. Originally archaea (Greek for "ancient things") were considered extremophiles, working away in salt flats, sulfuric hot springs and around deep-sea hydrothermal vents; now scientists are finding they are important in nitrogen and carbon cycles. Archaea are the organisms getting the bad rap around methane production in wetlands and in ruminant, protists and termite bellies. They also make up around 10% of the organisms in your gut.

Be deeply grateful for their contribution to your wellbeing. As your food passes through your gut system, it is the bacteria who produce most of the enzymes to break-down your meals. In this digestion process, they ferment the indigestible, to the digestible. They alchemise hormones and vitamins B and K. Bacteria affect your appetite and mood, teaming up with your cells to produce serotonin, the happy hormone. The gut microbiome influences human health and wellbeing far beyond the gut. New research shows that disruptions to our microbial partners (dysbiosis), is a contributing factor to, well, nearly everything to do with human wellbeing.

Take your pick of disorders from inflammatory, metabolic and autoimmune disorders, lung and heart diseases, skin problems, neurological issues and some cancers. Feeding our microbes a diverse diet, low in toxins and processed food, is essential to the wellbeing of all. Some scientists now argue that the human brain, may potentially be the central processing unit for the primary gut 'brain'.

This research into the human microbiome, is opening doors for our understanding of the vital role of bacteria in soils. From breaking down and fermenting organic materials, to solubilising rocks, immobilising (holding onto minerals), detoxifying, plant defence and synthesising vitamins, enzymes and plant growth hormones. In many ways, the relationship between bacteria and plants, mirrors our own gut (except for the bit about eating rocks). Bacteria are the soils clean-up crew. They feed upon root exudates (sugars and other materials released from roots) and dead organic materials; leaves, each-other and the poop and wastes excreted by larger organisms. The bacteria feed upon the less complex sugars, consider materials you mix into a compost to generate heat; this heat is due to bacterial metabolism. Composters call these simple sugars, the 'green' carbons: materials which include green grass, green leaves, molasses, sugar, kelp, manure and urine.

Bacteria make the first building bricks for our soil city structures, creating micro-aggregates. These are the very small crumbs. As with everything in life, balance is key. An imbalance, known as 'bacterial dominated soils', creates very fine, structure-less soils. These are more prone to erosion and surface crusting than soils which are higher in fungi. This leads to issues with air and water penetration. Bacterial soils also lead to nutrient immobilisation, which means nutrients are tied up and not freely available to the plant. Think of bacteria as a bag of fertiliser, closed and sitting in your shed. A healthy soil contains around 2.5 tonnes of bacteria per Ha (2200 lbs/ac). With 17% nitrogen in their bodies, this can equate to a tonne of urea in every hectare. Bacteria hold onto this nitrogen until higher microbes, nematodes or protists, consume them. The nitrogen cycle is then completed when the excess plant available N is excreted back into the root zone. It is common to find these bacterial conditions in dairy farms and cultivated fields, consequently, soil losses, weed pressures and the need for artificial inputs are high. Bacterial domination may be a major limiting factor on your place; these soils are the ones I call "constipated", or in the high rainfall regions; soils with "rampant diarrhea!" Bacteria are essential, unavoidable and key microbes in supporting optimal soil and plant health.

FUNGI

These are the organisms holding the fabric of soil life together. What sets fungi apart from the other kingdoms of life is chitin, a long chain carbohydrate found in their cell walls. As individual hyphae, they are invisible to the naked eye (2-10um wide), but as they adhere and bundle together, they become the visible mycelial structures you see on decaying wood or on your lawn as fairy rings. In the soil metropolis, the fungi provide the walls and mortar to build the cityscape. They are the neural network of the soil, sending information across thousands of acres, communicating insect attack, instigating plant defenses and sharing resources. Estimates point to at least 70,000 distinct species of fungi in soil across the world, which include yeasts, moulds and mushrooms. However, this figure is still an educated guess; many more still reside in the void. Soil fungi groups can be characterised by three main functions in soil: mutualists, saprophytes and pathogens.

As the term suggests, mutualistic fungi offer an array of benefits to plants and animals. There are examples of insects who choose to partner with, or even farm, fungi. These insects include leaf-cutter ants, termites, stingless bees and beetles. Lichen and algae require this mutualist relationship with many species of fungi for their survival. Mycorrhizae, or MF ('myco' meaning fungus, 'rhizae': roots) are mutualists, which live in-and-on plant roots; a symbiotically beneficial relationship between the plant, fungus and their bacterial partners. The mycorrhizae reduce plant stress, offer root protection and water and nutrient uptake in exchange for carbohydrates (carbon), energy and fatty acids. Their hyphae are much finer and can travel much further than plant roots, increasing the surface area for nutrient and water uptake up to 40X. They are a significant component of a healthy soil ecosystem, making up as much as third of the microbial biomass in soils.

Over 90% of plant families and 250,000 plant species have this MF relationship, supporting healthy, resilient plant growth. At the time of writing, there are 7 major mycorrhizal types; arbuscular, arbutoid, monotropoid, orchid, ecto-, ectendo- and ericoid. The endo-arbuscular mycorrhizal (AMF), (from the Greek word 'endon', meaning within), are types are fungi which live inside the cells of plant root extending their fine fingers out into the soil environment. Ecto-mycorrhizae (EcM), ecto- meaning they wrap themselves around the outside of the roots. Many trees and vines of agricultural significance are AMF, while some have AMF relationships early in life before switching to EcM.

Most plant species (85%) have an AMF relationship. This includes most of the leafy green plants and our agriculturally important crops, such as: lettuce, onion, peas, grapevines, almond, coffee, cherry, maples and cedars. The berry families except for lingonberries, blue-berries and cranberries are AMF. While nearly all flower and foliage plants and shrubs have this relationship, except for Azalea, Rhododendron, Banksia and Heath. Some cool season weedy grasses are less mycorrhizal dependent. Species such as Annual Bluegrass (*Poa annua*) can become less competitive with the promotion of AMF. Breeding selection has created commercial crops with less dependence on AMF, with consequences for the resilience of many of these species.

Around 10% of all plant species have EcM relationships, particularly seed producing forest trees. This group includes many hardwoods and conifers, such as Alders, Spruce, Oak, Chestnuts, Eucalyptus, Pine, Poplar and Willow. EcM species are far more diverse, over 6000, versus the hundreds of AMF. An individual tree may harbour a diverse array of hundreds of different EcM varieties. Despite only a small percentage of plants having an EcM relationship, they play a globally significant role, as the dominant temperate forest cover species.[11]

Without a host plant, most MF cannot survive, these are species that cannot produce their own fatty acids and are not saprophytic. There are some species, such as truffles, who can survive on decomposing organic materials, but they cannot complete their reproductive cycle without a plant host.

Around 10% of plant species are non-mycorrhizal species, including members of the Brassica family, like Canola, Lupins and Amaranthus and others such as Quinoa and Sugarbeet and the spinach family. Many common cropping weeds are non-mycorrhizal including Wild Radish, Kochia, Fat Hen and Pigroot. During the evolution of these plant families, the gene for symbiosis has become lost. Many plants that thrive in waterlogged conditions--such as reeds, rushes and Horsetail--do not need this relationship to flourish.

Saprophytic fungi, break down the more difficult to digest materials such as chitin (in fungi and insect bodies), cellulose and lignin materials. Unlike most other living things, fungi first digest their food outside of their bodies before ingesting them, using a range of acids and enzymes. These powerful acids can break chemical bonds, releasing minerals such as phosphorus, calcium, zinc and other tightly bound nutrients in the soil. In this breakdown process, they also release carbon dioxide, water and other by-products; some of these metabolites can remain in the surrounding soil for thousands of years under the right conditions.

The pathogenic or parasitic fungi on the other hand, steal from their hosts. These fungi can reduce plant production, or even result in plant death. Every year pathogenic fungi such as rusts, smuts and blasts and root rots, such as the aptly named "Take-all", reduce global crop volumes between 20-40%. Much of the scientific research has focused on this group, the "bad" fungi. Some pathogenic fungi may live in and on plant roots in a beneficial role to the plant, such as the entomopathogenic fungi (from the Greek entomo = insect, pathos = suffering), which derive their energy from the bodies of insects or nematodes. Mycorrhizae themselves, are not always beneficial. In certain conditions, they can also be parasitic; particularly when plant photosynthesis and root exudates are low. Human disturbed environments such as cultivation or soluble fertilisers, can favour MF species which are less beneficial or even parasitic.

In natural systems, there are no 'waste' materials. Fungi provide a vital buffer against the effects of many man-made pollutants. In a process termed "myco-remediation," they can transform even the most troublesome of our wastes, such as lead, cadmium, petroleum, radiation and organic pollutants in groundwater. Even the hangover from our modern world, plastic, can be broken down by oyster mushrooms.

Beneficial fungi are a sensitive bunch and are often the first to decline in modern farming systems. The balance in biomass between fungi and bacteria, is important for soil function, structure and plant health. On a planetary scale, the tiniest of its inhabitants are essential for the health and well-being of ecosystems, curbing nutrient leaching and sedimentation to waterways, and reducing greenhouse gas emissions to the atmosphere.

Yeasts fall into the fungal family, and when in balance provide many beneficial functions. They form beneficial relationships around roots, and on leaves can promote plant growth through the production of phytohormones.[12] Yeast has a beneficial relationship with B12 and cobalt. Even at low applications, yeast can measurably increase plant photosynthesis and production.

PROTISTS

The protists do not fall into the bacteria, animal, plant or fungal categories. These microscopic organisms are loosely grouped by the presence of a nucleus and membrane. In evolutionary terms, these organisms represent the original ancestors of all living cells; your great, great, great...grandma or pa, which morphed out the depths around 2 billion years ago as oxygen levels rose.

In a teaspoon of healthy vibrant soils, there may be as many as a million of these creatures building the corridors, living spaces and highways through the soil environment. Single celled and mobile, protists have diverse and vital roles in both soil and plant health. They speed up nutrient cycling, control bacterial populations and stimulate plant defence.

There are 3 major groups in soil: flagellates, amoebae and ciliates. Flagellates are the smallest, named for the one or 2 whip-like flagella they use to zip around soil. Many of the amoebae are much slower moving, using a pseudopod, or false foot, to ooze through the soil moisture, absorbing all who fall into their path. Amoebae can consume a variety of food sources, some herbivorous species resemble tiny bloated dairy cows, peacefully grazing on algae and dead organic materials. Testate amoebae, who fashion themselves in hardened shells for protection, have been observed hunting together as a pack, to track down and attack bacteria-feeding nematodes. While the 'vampire amoebae' (*Vampyrella*) drill into algae and fungi and then suck away their life essence. The larger ciliates (up to 500um), named for the hair-like cilia which cover their bodies, zip through soil feeding upon bacteria, fungi, other protists and nematodes. Their bodies also provide tasty morsels for hungry micro-arthropods.

Protists are one of the main groups responsible for nutrient cycling to the plant. These voracious animals may eat as much as 10,000 bacteria in a single day, excreting plant-available nutrients in their wake. As they feed upon bacteria and fungi, in a dynamic relationship with the plant, they selectively graze upon bacteria or fungi to maintain the bacteria: fungi (B: F) ratios in the soil. As they are important predators for bacteria, a deficit in protists can lead to a boom in bacterial populations, creating a cascade of impacts throughout the soil foodweb.

All the microbes in the soil foodweb are interconnected, in part why it's called a "web." The term first arose in the 1960s with the realisation that all organisms are limited by food and are connected to each other. These relationships are often complex and unpredictable. Take protists for instance, in the presence of disease organisms, they stimulate bacteria to produce signaling molecules, which aid the plant in self-defense.

In a process called a "microbial loop", plants create conditions more conducive for growth. By releasing root exudates, plants stimulate bacterial populations, which in turn attracts more protists and nematodes. These predators then release the nutrients bound in the bodies of bacteria in plant available forms back to the plant.

NEMATODES

These are the non-segmented worms and are colossal beasts compared to the bacteria, ranging in length from 300um to the just visible size of 2.5mm. Most research and attention focusses on the "bad" guys, the root feeders and the parasites, which make up perhaps 5% of all nematode species. When soil systems are out of balance, the damage from nematodes can be considerable; just in New Zealand, the costs from pasture yield losses, are estimated to be over $1 billion NZD a year. As with all microbes, there are no "good" or "bad" guys, just indicators and cleaning crews doing their work.

Most nematode families are beneficial and have important roles in building soil structure, the city highways and byways. They control and balance microbial populations and drive the N-cycle. They live in areas where their food concentrates, around plant roots and in organic matter. Nematode abundance in soils, can range from 1 to 10 million individuals in a square meter of soil.[13] You'll find nematodes in every environment around the world; on the ocean floor, the top of the highest mountains, ice sheets in Antarctica, in the hottest deserts and working away under the snow in Montana. Most nematodes make their home in the top 15cm of soil; however, they have been found as deep as 1000m, in a gold mine south-west of Johannesburg.

The 40,000 different species of nematodes are grouped by what, or who, they eat. Some peacefully graze upon algae and organic matter, fungi, or feed more voraciously upon bacteria, roots or each-other. The structure of their mouth parts is an indicator for who or what they feed upon. Some nematodes alter their food of choice as they mature, starting out life as bacterial feeders before moving onto larger prey such as the protists and other soil nematodes. While the more adaptable omnivorous nematodes can feed on different foods depending on what foods are available and the conditions in the soil.

Specific nematode groups are canaries in the mines of soil health. Missing key groups of nematodes, has been proposed as a method for assessing soil integrity and health, as they are more vulnerable to disturbance and pollution. These species are often predatory or omnivorous, larger and longer-lived, compared to the faster reproducing bacterial feeders.

When nematode levels are low, there is an associated reduction in plant nitrogen, disease and pest suppression, soil structure development and root health.

MICRO-ARTHROPODS & MICRO-ANIMALS

This group contains the demolition crews, charting new territory and building new channels for air and water. They also provide essential services, collecting garbage and recycling materials. As the name 'micro' suggests, you'll need a microscope, or in some cases just a small magnifying loupe (X10-20) to view them. Arthropods have segmented bodies. These are the: micro-bugs; the insects, centipedes, mites and spiders. They feed on a wide-range of materials, from organic matter, lichen, bacteria, fungi and each-other. They also provide an essential food source for the larger bugs. They shred and digest organic materials, speeding up decomposition, while their poop inoculates soils with bacterial and fungal spores.

The arthropods are nitrogen thieves, accumulating N up the food chain as ammonium. Their bodies contain 7-14% nitrogen (non-legume plants only contain 2-4%), upon their death their bodies make up a critical fraction of the available nitrogen pool. Many arthropods have a mutualistic relationship with N-fixing bacteria in their guts, which can contribute up to 40 Kg/Ha/Yr in healthy ecosystems. With insect populations in serious decline across the planet, we cannot afford to lose their valuable, and historically unaccounted for, contributions to healthy food systems.

The micro-animals include rotifers and tardigrade. The tardigrade, also known as moss piglet, which yes, is as cute as it sounds, bumbles through soils with its 8 stumpy legs. This miniscule animal (up to 0.5mm in size) is virtually indestructible; surviving in conditions above boiling and below absolute zero. They can withstand 10,000 times the gamma radiation that would kill you. They have mastered the art of cryogenesis, desiccating their bodies until conditions are more ideal. Tardigrades have been on the planet eating algae and occasionally each-other, for at least 530 million years, bumbling through five planetary mass extinctions. Scientists even sent a team of tardigrades into space to see if they would survive. The answer was yes, but just barely and they weren't terribly happy about the experience. The only place on the planet you won't find the tardigrade? ...where there has been a long history of cultivation and herbicides.

The transformative shifts regenerative producers witness on the land, have been difficult to rationalize, based on current agronomical thinking. We've been 'tickling' the system with products, that in many cases, would be the equivalent of parts per billion and seeing plant and soil responses. Again, and again, we find that by subtly stimulating biology, or addressing enabling factors, responses can be far beyond what seems rationally possible.

That is until you consider the dynamic of quorum sensing (QS). Quorum sensing is a microbial community's ability to sense and respond to signals. The process of QS was first discovered in bio-luminescent bacteria in the 1960s. With growing concerns around the exploitation of antibiotics, medical research in the '90s identified QS's role in streptococcus virulence. At any one time, you may have some strep or candida organisms in your body without any complications. When populations are small, microorganisms are unable to express certain genes. Gene expression takes a community. When numbers increase, the full gambit of gene expression can now switch on and you end up with a sore throat or a nasty itch.

In a process termed quorum quenching, signals are used to inhibit invasion or growth of another organism. A QS protein has been identified which communicates to cholera parasites to signal that their population has grown too high, in response the cholera leaves their human host. This signal will revolutionise cholera treatments and most interestingly, is also the same protein signal used to turn off fungal diseases in plants. QS signals have also been found to explain the efficacy of many traditional Chinese medicines, taking the lid off for new discoveries with countless medicinal herbs and mushrooms. If medical science can replicate these signals, they could create a range of new medicines based on biomimicry, to combat many common diseases. This signaling has opened the doors to potential new healthcare methods. These chemical signals can replace the broad scattershot approach of antibiotics, potentially diverting the feared bacterial resistance apocalypse.

In soils, we are only just beginning to become aware of the significance of these signals. The soil is saturated with thousands of metabolites, which create a sophisticated signaling network between plants, roots and microbes. These communication signals including hormones, organic acids, acyl-homoserine lactones (AHLs), peptides and other proteins. These signals enable organisms to identify each other, hide from or block attackers, or even camouflage themselves to sneak up on an unwary protist or plant root.

Fungi can intercept the hormonal signals being sent by nematodes to ensnare them. The plant responds to signals, produced by microbes, to change root exudates, induce defenses, increase root formation, branching and elongation, lift transpiration and metabolism.[14] These signals contribute to the organic matter matrix, made up of carbon, proteins and organic acids.

In soil, QS may lead to virulent disease or the switching on of the genes responsible for immunity and defence, biofilms, drought resilience and plant growth promotion. Dr Gwen Grelet, a vivacious soil molecular ecologist based at Landcare Research in New Zealand, believes that deepening our understanding of how these organisms communicate to each other, has huge value to producers. "Through understanding a little of the complexity, we can adopt methods which encourage the communication, supporting optimal health and integrity."

Dr David Johnson, at Chico State University, is showing that the QS signal proteins, the regulation messengers, are up to 10 times higher in healthy soils than stressed ones. These signals correlated with an increase in the 'housekeeping genes," the genes for drought resistance, plant growth and regulation, nutrient increases and substantial increases in soil carbon use efficiencies. Around 60-80% of the human immune function, is due to gut microbial actions; Dr Johnson suspects the same holds true for soil and plant health.

Soil truly is the new frontier for exciting breakthroughs and discoveries. Through deepening our understandings of how we can feed, support and harness our micro-herd, we hold the key to creating regenerative, productive and profitable landscapes.

5

First there was light

"In less than a hundred years, we have found a new way to think of ourselves. From sitting at the center of the universe, we now find ourselves orbiting an average-sized sun, which is just one of millions of stars in our own Milky Way galaxy." Stephen Hawking

My one impossible dream would be to pilot a time machine; to observe how landscapes, plants and animals flourished, before modern human hands molded what we see today. Setting the dial back 420 million years to the beginning of the Devonian period, landscapes would be unrecognizable, more like the landscape on Mars, overlaid with raw materials, such as sand, silt and clay. Enter the golden age of fungi!

When you think of fungi, what may come to mind is their sexual organs; that's ok, most people think that way. The showy toadstools and mushrooms are the eye-catching organs many fungi use to spread their spores. However, fungi spend most of their life as hyphae, microscopic threads. Building a clear picture of prehistoric life for these delicate organisms is, understandably, problematic. When paleontologists first found fossil evidence of these giant fungal sexual organs, some reaching 8m (26 feet) high, they were so unusual, it took 150 years for any scientific agreement that these fossils were in fact fungi. These fungal rulers were conducting a planetary mining operation, releasing minerals off rocks using a variety of acids and other exudates, weaving the fabric of life together. A role they still hold today.

While these fungi were enjoying their lavish feast of minerals, most plants were lounging in the seas, drawing ample energy from the sun. The sea was a far more hospitable environment, rich with minerals, more stable temperatures and ample water. At this time, only a few vascular plants had made their way out of the sea, restricted to life close to watery bogs.

How did plants move onto land, or why would they even want to? Scientists speculate that changes in sea levels, possibly because of tectonic forces or climatic changes, exposed sea plants to new environments. These plants needed to withstand scorching temperatures, desiccation and salinity. These were skills their cousins, the lichens, were already masters in. Lichen are a genius partnership between cyanobacteria, fungi and algae. Algae provide sugars from the sun, while the fungi reimburse them with water, minerals and protection. Researchers now believe this liaison provided the steppingstone for plants to move out of the sea and onto land. The giant fungi struck a deal with more basic plant forms- the bryophytes and algal mats present in the early Devonian soup. Fungi began to farm plants for their rich source of energy (sugar) and fatty acids. In return, the fungi provided plants with life-giving water and soil-derived nutrients. Fungi no longer had to invest in massive phallic structures to reproduce, leaving the costly process of obtaining energy and growing towards the sky, to their plant companions instead.

Developing terrestrial root systems and partnering with mycorrhizal fungi, led to plants transforming the skin of the planet. Mineralisation of rocks and the drawdown of carbon into the ground, altered global climatic conditions, to produce the cozy conditions for life as we know it. This fungal friendship literally mushroomed plant success, representing one of the most important relationships on earth. Really. Without it, you would never have been a glint in your mother's eye. If you are interested in yield, crop quality, pest control, water quality, reducing soil losses and greenhouse gas emissions, or building resilience to climatic pressures, then it's time to get interested in fungi!

Mycorrhizae can promote the growth of beneficial pest predators and in some cases, be predators themselves! In 2001, researchers from the University of Guelph in Ontario, were looking to see how much mycorrhizae springtails were eating. Springtails are tiny, grey-white, insect-looking critters who love to eat fungi. They tried several plant species and their mycorrhizae, which the springtails consumed with glee—until they put springtails around the roots of eastern white pine.

Their findings were unexpected. John Klironomos, the study leader, discovered that instead of the insects eating the fungus, the fungus ate the springtails! It "was as shocking as putting a pizza in front of a person and having the pizza eat the person instead." The researchers calculated the springtail slaying mycorrhizae were absorbing nitrogen from their tiny bodies, enough in fact to supply the tree with 25% of its annual nitrogen needs. The synergies of nature are complex and exciting. And I love when research doesn't go the way it's predicted.

For Geoff Thorpe at Riversun Nurseries, quality and soil health are non-negotiable number one factors for success. Grapevines are highly mycorrhizal, with a variety of endo-arbuscular mycorrhizal (AMF) relationships. Vines with poor colonization will grow stunted, suffering from nutrient deficiencies. Without AMF, they come under increased risk from pest and disease attack. As the hyphal strands weave through the soil matrix, they secrete an enticing range of materials; acids, enzymes, metal ions, amino acids, proteins (nitrogen) and sugars. An increase in these materials increases plant health and resilience, allowing a farm or ranch to recover quickly after climatic shocks.

To discover if plants have good colonisation, root samples can be sent to a soil microbial lab. Riversun partnered with Microbe Labs in Australia, to set up their PLFA testing facility on-site. Many organisms produce specific or signature PLFA biomarkers, which can then provide an accurate fingerprint of soil microbes present and active in a soil. Testing can also quantify proteins such as glomalin. Both glomalin and fungi contribute to the large crumbs or macro-aggregates in soil. These crumbs have a vital role in soil, holding much more carbon and water than their finer counterparts. This structure also protects valuable carbon from being exposed to the air and voracious bacteria.

As part of Riversun Nursery's vision, they consider the four elements--earth, air, fire and water-- as the foundation for decision-making processes. The aspect "earth" places a strong focus on building soil structure and organic matter. The nursery has been able to maintain their OM~10%. This is extraordinary after 14 years of growing vines using the 2 year on, one year resting and green cropping strategy. The benefit from their soil and climate enables them to grow three successive green crops on the same land in one year. Extensive research shows that most of the world's intensively managed agricultural lands lack adequate levels of MF. When thinking about what practices would disrupt hyphal strands and undermine aggregation, the first thing that comes to mind is cultivation. Discs, rototillers and moleboard ploughs destroy soil structure, reduce MF colonisation, glomalin and the pool of MF spores (propagules).[15] Unfortunately for Riversun, the nature of rootstock planting requires intensive soil disturbances; soil beds are prepared to a fine tilth and at harvest, the soil is ripped up to extract the vines. To mitigate soil and carbon losses and disturbance to the microbial community, they now rely on the multiple green crops grown in the resting year to replenish the soil microbiome. These cover crops are incorporated pre-flowering into the soil, to provide organic matter, increase MF spores and feed microbes.

To ensure a steady supply of MF spores, Riversun has planted over 3 km of permanent rows of alley crop species. These plants are MF hosts, such as rosemary and lavender. They bring other benefits, including pollinator feed and wind breaks between the blocks, while wafting a sweet perfume across the nursery. Ensuring high MF colonization is essential to produce a more uniform rooting system, increase plant defence and provide a valuable aid in addressing climatic uncertainty.

Mycorrhizae are a vital part of the water story, providing drought protection for most plant species. With their fine hyphae, MF can probe deep inside the soil crumbs, accessing pools of water unavailable to plants with their thicker, shorter root hairs. Fungi also absorb water when there is adequate soil moisture, which they slowly release during periods of drought. Not only that, fungi also make water! As they decompose carbon materials (CHO) up to 20% of this volume is converted into H_2O. With Riversun's focus on a goal of 1000 year sustainability, they view water as one of the most formidable issues facing production. The 'water' element at Riversun is expressed through large re-vegetation projects and dams to store and slow water movement and through increasing water storage directly in soil (high OM and good soil structure). Enhancing MF to improve water use efficiency is a tool readily available to today's food producers.

Endo-mycorrhizae which are found on grasses and many tree species are invisible to the naked eye. Without a laboratory or a microscope to view hyphae, low MF may be inferred by healthy non-mycorrhizal species (*Chenopodium, Brassica* etc.), poor soil aggregation (crumbs) and 'clean' roots. In the field, poor MF colonisation shows up through increasing vulnerability to drought, slow recovery after grazing, low P and Zn in the plant, root diseases and increasing weed pressures. Sending root samples away to biological laboratories will provide you insights on the situation on your own property. If you're already set up with a dissecting microscope, you can do your own stains and testing.

Mycorrhizae make delicate structures that are adversely affected by certain pesticides, chemical fertilisers (especially soluble phosphate), extensive cultivation, organic matter loss and erosion.[16] Insect pests can also reduce MF. Glyphosate, the world's most popular herbicide can reduce the colonisation of mycorrhizae by a quarter in the following season's crop.

To put it bluntly, most of our modern farming practices undermine the soil's gut function and disrupt the most valuable soil microbes we need for quality food production, as well as the most important organisms to ensure we can continue to produce food. Mycorrhizal growth and repair is supported through cover crops, carbon inputs, pasture cropping, compost/vermicast extracts, cell grazing, reduced soluble fertiliser, particularly phosphorus inputs and the sparing rotations of non-mycorrhizal plant species (see appendix). There are many commercial inoculums and easy-to-use products that can now help speed up MF repair. Keep in mind that management is always your number one tool to restoring MF function. First, identify what's causing the poor colonisation. Often, it's our actions which have been undermining this vital partner. A first step may require getting out of nature's way and stopping the actions which kill one of the planets most beneficial organisms.

To increase mycorrhizal colonisation and activity:

* Firstly, stop the actions which are killing them!
* Remove chemical 'i-cides. Buffer herbicides with fulvic acid and avoid high soluble phosphate.
* Increase plant diversity.
* Inter-crop alley crops between cash crops. There are cover crop species which are rich sources of AMF spores: Flax, Sorghum, Millet, Sudan Grass, Sunflowers and Oats. Even trees, which may have ecto-mycorrhizae, have a stimulatory relationship with the AMF in grasses.
* Use carbon-based inputs and biostimulants to encourage underground diversity.
* The presence of plant growth promoting bacteria (PGP) partners such as yeasts, pseudomonas fluorescens and bacillus species work in partnership to increase the efficacy of MF.

If mycorrhizae are critically low:

Stimulate and feed MF with soluble humates, compost extracts or vermicast. If you are planning to apply herbicides or cultivate a field, then the addition of soluble humic or fulvic products is a must to support your beneficial MF populations. Mix fulvic with herbicides, or drip onto soil when cultivating. Do not miss an opportunity to feed your essential micro-herd!

* *You can make a mycorrhizal inoculant* – combine potting mixes with soil/leaf materials collected from local healthy ecosystems. *Grow AMF spores in soil using C4 grasses, such as Sudan grass, Paspalum spp, Corn etc. See appendix for instructions.*

Bio-fertiliser markets are booming in response to producer demands around the world. In 2016, the market was valued at $1.1 billion USD and is projected to grow by 14% every year.[17] These biofertilisers include nitrogen fixers: rhizobia, mycorrhizae, azotobacter, azospirillum and phosphate solubilising bacteria.

Ten years ago, I heard from two growers who both added a mycorrhizal product to increase corn yields. One grower saw a 25% yield increase and the other saw no measurable difference. On deeper examination, the grower with the yield increase had critically low AMF in the fields not treated. The one who saw no response already had good natural AMF levels.

If you're going to take on these products it is important to check what type of propagules you're buying, as there are considerable differences in their effectiveness. MF products may include spores, colonized root pieces and dried mycelium. The root pieces and mycelium may only be viable for a few weeks, whilst spores may survive for years. Spores require much longer timelines to colonise roots. This needs to be factored in as a laboratory may use viability test protocols based on inocula from root materials. The challenge for producers using these products is assurance that what's in the bag, is in fact alive and will go to work in your soil. Independent testing in New Zealand in 2011 revealed that out of 7 commercially available mycorrhizal products, none were viable. A wake-up for suppliers and producers! As it can take 4-6 weeks for colonization, many universities and research centres that provide viability testing services with a quick turnaround may be hastily jumping to a premature conclusion that products are non-viable. So, it's prudent to ask the labs what their protocol is and demand viability testing from any suppliers when buying biological inoculums.

Personally, I would rather address why MF is low in the first place and then go to work on increasing native populations. Spores can remain in the soil for many years and very few producers have zero colonization. While it's not unknown, the few cases where I've seen this have been associated with high disturbances; poor irrigation management, heavy chemical use and following canola or commercial forestry removal. Those indicators lead straight back to the earlier discussion on looking at "enabling factors."

As you can see, plants do not exist in isolation. Ponder this for a moment, many plant vital functions are external to their body; they outsource essential functions to microbes that are responsible for immunity, nutrient and water availability. Their gut, kidneys and thermostat are outside of their bodies. If you consider the primary goal of soil management, is to support an optimal digestive system, it is the fungi that supply powerful gut acids, vitamins, enzymes and minerals to fuel energy and health. Can you grow plants without soil and biology? Yes. Will they be healthy, have full genomic expression and be nutrient dense? No.

Soil carbon (C) has become increasingly politicised – mostly in conversations around key greenhouse gases; methane, nitrous oxide and carbon dioxide (with very little talk around the other major climate influencer; water vapour). These various solutions and schemes are making carbon quite the political talking point. Why is it that everyone is getting so hot and bothered about carbon? Universally, it's a similar mechanism behind our addiction to petroleum. Carbon fuels every aspect of our lives, from the gas in your car to plastic, fertiliser production, the heat in your house and electricity generation. Carbon is stored energy. Stored energy in the atmosphere holds more heat and stored energy in soil provides fuel for microbes and plants. It's a measure of the engine room of your soil and ultimately your profitability.

All organic living materials contain carbon. It's the very elixir of life. There is a finite amount of carbon on the planet; every single carbon atom has passed through a living being at one point. Your body is included, having grown from atoms which were once in the bodies of dinosaurs and trees. I love the idea that at one time the carbon molecules inside me were once the brain of a megalodon or T-Rex!

Soils contain a massive reserve of carbon, larger than the combined pools held in plants and the atmosphere, containing around 2,344 billion tonnes of organic carbon. Since the very first sod of soil was tilled to plant a crop, the global losses of C began, with losses from land use estimated to be between 133 to 200 billion tonnes. That's around a fifth of all C in the atmosphere. Let that sink in for a minute. There is a catastrophic event happening across the planet which society has been blind to. We've been so focused on fossil fuels and the short-term methane from burping cows that we've missed the significant amount of carbon (and water) being lost from under our feet to the air and to the seas.

The news is not all bad, though. If carbon can be lost on such a scale, we also have the means to rebuild it. Soil C is the most immediate and cost-effective tool in addressing the legacy load of atmospheric carbon.

What is the Soil Carbon Pool?

The different components of soil organic matter (SOM) all include carbon; the living biomass, plant residues, humus, dissolved organic carbon, recalcitrant organic carbon and inorganic carbon. Soil organic matter has a crucial role in the physical, chemical and biological function of soil as well, providing the soil's valuable energy storage system. SOM contributes to nutrient retention and turnover, soil structure, moisture retention/availability, degradation of pollutants, carbon sequestration and soil resilience. Soil organic carbon (SOC) is chemically and biologically active, decomposing relatively quickly (years to decades).

SOC is estimated to be around 58% of SOM; if your laboratory measures SOM, divide the lab result by 1.72.

Living microbial biomass includes all soil organisms large and small: from earthworms, termites, beetles, to bacteria. They all contain nitrogen, carbon and water, which is held in their bodies and released upon their deaths. They also excrete materials concentrated in carbon--poop, wee, spit and vomit, essential elements in the 'stickiness' that creates stable healthy soil.

Plant residues--made up of roots, shoots and leaves found in-and-on soil. Depending on the soil depth, biological activity and the amount of cellulose/lignin, these materials can be broken down relatively quickly (from days to decades). These residues provide a vital energy source for microbes, which is respired as bio-available carbon back to the plant.

Particulate organic carbon (POC) includes materials larger than 2mm, such as roots, dead insects and leaf litter. Dissolved organic carbon (DOC) or the 'litter tea' is found in the soil solution and is readily available. It can turnover in minutes or weeks. It represents a small, yet vital fraction of SOC. The POC and DOC pools are called 'labile carbon' as they cycle relatively quickly in the soil.

Humification describes the biological process which changes organic matter-for example, roots, leaves, manure, or a dead sheep-into the fully decomposed dark uniform material known as humus. Humus is made up of organic materials less than 0.053 mm in size. It is the end-product of organic material breakdown which includes glomalin, amino acids, proteins (nitrogen), enzymes, lipids, fats, vitamins, growth factors and other signaling metabolites. It is difficult to define by its chemical or biological make-up, so difficult, that some scientists are calling for the term to be removed from scientific terminologies.

Humus has a complex structure which acts as a storehouse for minerals and water and provides an important energy source. This carbon pool is more resistant to decomposition by soil microorganisms and so tends to turn over more slowly (over decades to centuries). It plays a role in all key soil functions and is particularly important in the provision of nutrients.

Increasing carbon in soil creates a sponge, improving nutrient cycles, water holding and quality, requiring less need for inputs and artificial props. This is why regenerative agriculture places such a key importance on soil C.

Measuring soil carbon accurately has been a little like trying to catch mist in your hand. Every day, the soil breathes carbon in and out. In my opinion, the scientific community's inability to agree on measurement has been an active agent in slowing down innovation and action. These scientific disagreements have (unintentionally or not) slowed the uptake of regenerative practices throughout the wider agricultural community. Carbon is hard to accurately measure, how to measure C is the wrong question to be asking. Carbon is tied into every aspect of healthy functioning soil, aggregation, structure, water, nutrients and gas diffusion.

I stumbled across an article a few years ago about an inventor looking for investment to fund CO_2 scrubbers. These machines could remove carbon from the air and store it deep into soil.... Exactly what plants already do. Nature figured this one out 3.4 billion years ago, when the first photosynthetic bacteria discovered the joyful excitement contained in sunlight energy. Over the past decade, there has been a growing focus from industry and researchers, fired up to crack the chemical wizardry of photosynthesis. Gene-editing techniques are now being developed in the field to tweak the enzyme responsible for photorespiration. Researchers call photosynthesis "slow and confused."[18] Yet, all our attempts to imitate nature thus far, have been incredibly expensive, wasteful and a pitiful imitation. We are still grasping in the dark to try to replicate the efficiency of nature's systems; focusing on mechanical and technological solutions for a biological problem already solved.

Think back to your teenage years at school when you were likely taught about photosynthesis; perhaps your eyes glazed over and you wondered what this had to do with your life? At least that's my memory. Photosynthesis, however, is the most important process on your land, for the planet and one you want to get more familiar with. It governs the resilience, health and, ultimately, profitability. During the process of photosynthesis, solar light energy penetrates the leaf surface. This excites a chain reaction, providing the spark of energy required to cleave water into its component parts: electrons, hydrogen and oxygen atoms. Breathing in carbon dioxide provides the building blocks for life, combined with hydrogen and oxygen, carbohydrates (CHO) are formed. The oxygen respired out, is what enables us to live upon this planet, while CHO convert into sucrose, starch, proteins, lipids, cellulose...literally everything with an organic base. Everything you produce on the land starts as plant sugars.

All plants photosynthesise, that's a given. What is not assured, is the efficiency of this process, which scientists estimate may range from as little as 0.1 to 4.6% capture. Increases in microbial biomass and diversity, soil aeration, water availability, carbon: nitrogen ratio (C: N) and the availability of calcium and other minerals can all enhance this efficiency. Photosynthesis requires a range of complex and interactive mechanisms many of which involve essential mineral catalysts, such as phosphorus, nitrogen and magnesium to provide the spark.

There are two major carbon cycles at work in soil: the one most studied, is the short-term decomposition cycle, or the one I call the: "I'm here for a good time, not a long time, honey" cycle. This essential process involves the addition of organic materials, like roots, dead leaves, compost, manure and urine. These materials fuel the diverse community of soil microbes, worms and insects that either respire or poop excess carbon back into the soil. This activity mostly concentrates in the aerated top six inches of soil. As they toil away, day and night, carbon breathes in and out of the soil, powering up plant photosynthesis. This carbon breath is in flux, influenced by temperature, moisture, time of the day/year and food availability for microbes. This cycle plays a critical role in providing carbon back to the plant for photosynthesis. As biological communities advance from more primitive bacterial conditions towards higher fungal biomass or activity, then this atmospheric out-breath may reduce to 25%.[19, 20]

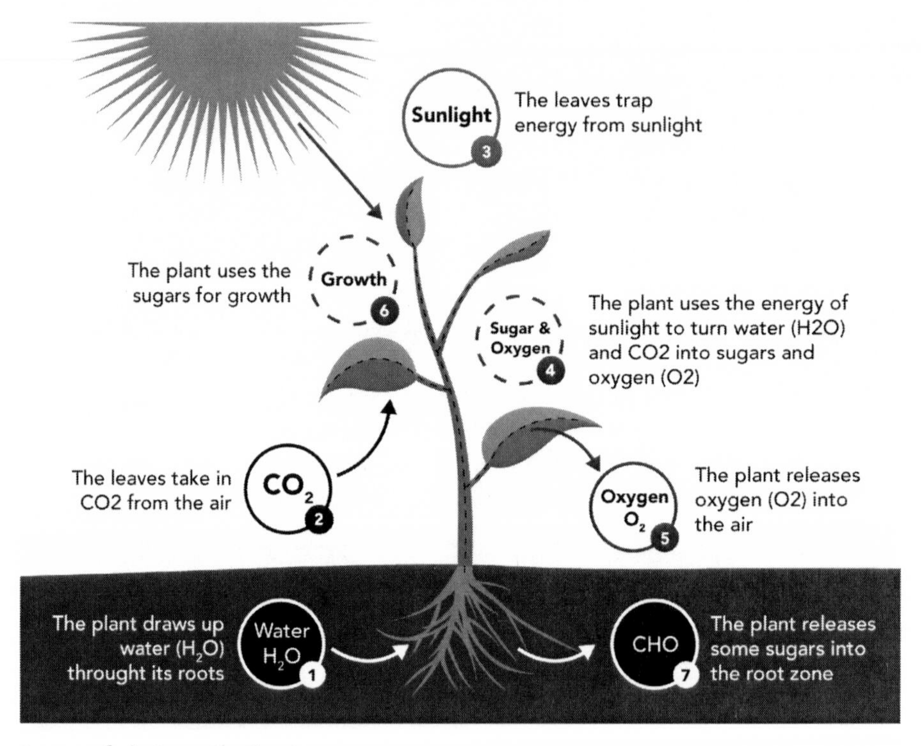

Image of photosynthetic process.

The second major carbon cycle is the one less studied; the deep drawdown of carbon into soil. This is the carbon captured during photosynthesis and delivered to the soil via plant root exudates. In advanced grass species, over half of these valuable sugars from the photosynthetic process end up being released out the roots. For the soil microbes this is what nectar is to airborne insects. These exudates contain an enticing range of enzymes, fatty acids, growth factors, vitamins and quorum signaling agents- signals which effectively either wake up, ramp up, slow down or put microbes to sleep.

Much of this root nectar is held at deeper undisturbed levels in the soil, 150 - 400mm (6-25"), deposited there through the action of our friends, the mycorrhizal fungi. Coined the "liquid carbon pathway" by Australian grassland ecologist Dr Christine Jones, this deeper soil carbon may remain chemically or biologically protected in soils for thousands of years. When exposed to the air and water, these fungal carbon deposits, including glomalin, chitin (the skeleton bodies of dead fungi) and melanin, are more protected from degradation than the root derived sugars.

Root exudates are the cheapest, most-efficient and most beneficial form of organic carbon for soil life. Exudates and microbes concentrate around the root zone, or rhizosheath, creating visual signs we term the "Rastafarian roots." These root systems resemble dirty dreadlocks, giving plants vital protection against changes in pH, aluminum, salinity and temperature or moisture fluctuations in the environment.

Herbicides, fungicides and overgrazing have been shown to shut down this important soil process. In the field, the excessive use of soluble nitrogen and phosphorus forms, (synthetic or from manure) is effective at driving plant production, often to the detriment of plant photosynthesis. As amino acid nitrogen levels lift in the plant, the products of photosynthesis decrease, and insect pressures increase. It's these carbon to nitrogen (C: N) ratios in the plant which lie the heart of producing nutrient dense food with integrity. C:N ratios influence yield, flowering and responses to environmental stresses: salt, temperature, drought and nutrient deficiencies.

While fungi can excrete acids to extract minerals from rocks, they can also create and transform rocks, in a process called geomycology. Fungi literally form new rock minerals and crystals, such as weddelite, calcium oxalate and glushinskite. These minerals incorporate carbon into inorganic forms, meaning fungi play a major role in effectively sequestering carbon over eons. Dr. Richard Teague, a range ecologist with Texas A&M has been measuring lifts in this "inorganic C" under adaptive grazing systems in Texas, USA and Alberta, Canada. This is a newly developing field of research, which may hold a critical key to understanding carbon dynamics and the important role microbially diverse ecosystems play in global carbon cycles.

Soil carbon cannot be considered in isolation, it is intimately linked to nitrogen dynamics, water cycles, soil stability, resilience and food quality. The benefits offered to producers, society and the environment are immense and broad reaching. When you consider what historic carbon levels existed in many landscapes before agriculture, it becomes clear what an incredible loss has occurred. Soil Scientist Rattan Lal proposes that 50 to 70% of this carbon has been lost through management. This means that across the planet, we have effectively washed away 50- 70% of our water holding potential.

So how can you tell if your soil is losing or gaining carbon? One way is to take a soil test, which gives you a small part of the picture or take a deep core which will show carbon levels at depth. This may not be informative if you don't have data for comparison, however. There are also labs that can test for glomalin; the carbon by-product produced during this biological process. These are the results which will give you a clear picture of whether your management practices are building your soil resource or degrading it over time.

A cheap and quick method is to dig a few holes and compare the colour of your topsoil, to a hole dug in a nearby undisturbed area which hasn't received fertiliser, been cut for hay, cultivated, or been intensively grazed. If you see a visual difference and your soil is paler, this can show management changes are required. Apply a little water to the soil surfaces, as these colour differences can be clearer with moisture present.

Can you have too much carbon? Ask someone with thatch, peat, or muskeg soils.

For decades regenerative producers have used a simple, yet controversial tool called a refractometer or Brix meter. Refractometers measure the carbohydrates and total dissolved solids contained in plant sap. Commonly used in vineyards and orchards to assess fruit ripeness, they can also be used to assess plant sap and photosynthesis. It's an easy-to-use handheld tool, which provides instant feedback about crop health. See the appendix for insights into how to use a refractometer.

A visit with most regenerative producers typically involves a discussion around their progress with Brix and their joy when lifting Brix levels into the teens and above. Simon Osborne, a seed and crop producer in Canterbury, New Zealand, keeps a beat-up Brix meter wedged beside his truck's handbrake. The Osborne farm covers over 270 hectares (670 acres) of predominantly heavy clay with some silt soils. He also runs an arable seed business with sheep in the inter-cropping phase. In 1976, his father was one of the first to adopt "no-till" practices, replacing cultivation practices with glyphosate applications. Initially there were some soil structure improvements, but minimal increases in carbon. When Simon returned with his wife Angela to farm full-time in 1992, he adopted Conservation Agriculture (CA) practices, "I decided to reboot the system, stopped burning crop residue, instead shredding and leaving residues on the surface." He is quietly proud of the soils he has developed, with soil carbon levels gradually rising from 1.5%, to an average of 5% C (9% OM in the top 150mm (6")).

Like all of the best regenerative caretakers I've met, Simon has excellent observation and action skills. Brix has been slowly lifting and readings are now consistently in their teens. Eliminating fungicides and pesticides has enabled his soil health programme to progress and yield has still been maintained satisfactorily.

Sap Brix is a real-world measurement, not confined to disrupted microbial systems or pot trials in greenhouses. In the field, Regenerators commonly find that as Brix comes up, so too does biological activity and mineral availability. An increase in Brix levels corresponds to a lift in crop health and resistance to frost, pests and disease. As Brix levels lift, so too will the deep soil carbon (C), improving water and nutrient cycles. Brix is directly related to farm profit, with a lift in milk production and weight gains. American grazier and Regenerator, Dr. Allan Williams, has found that every 1.0% increase in Brix adds 0.1 to 0.3 pounds in average daily weight gains. In dairy we've found this equates to around 100 gms of milk solids (MS) per cow/day.

Making claims that Brix relates to crop health is hugely controversial, at least in some circles. Much of this dispute arises from conventional agriculture proponents and a lack of supporting scientific literature. Scientists and rural blogs state that sap Brix levels are not a reliable measure. I do not believe it is the measure that has failed here--more the linear mindset applied to the study design and implementation. With low humus, poor biological activities and soluble fertiliser inputs, Brix levels will fluctuate wildly and are unreliable.

It's important that you use this tool to trend your own success, not to compare to others. If you do not see a trend of increasing plant sap Brix levels over time, then, potentially you're not regenerating your soil. It's time to return to the 5 M's and investigate what is slowing your progress down.

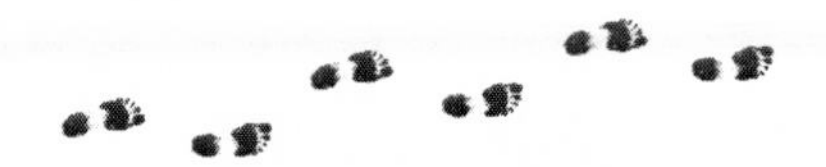

Dr. David Johnson's foray into the relationships between biological biomass and yield, began in 2010. Together with his astute wife Hui-Chan Su, we all had the opportunity to geek-out during a conference in Albuquerque. The following day offered up a free morning and time to explore some of the remarkable etched petroglyphs the area is famed for. Typical for most scientists, we became so absorbed in rocks and history, that I nearly missed my lunchtime flight. My immediate impression of David was his attentive listening skills. In the past, the notion I get from many cutting-edge scientists, has been that they can dominate conversations in their passion to share their revelations; a little like they're delivering a lecture to a large hall. With David however, I found vivid blues eyes focusing intently, his head cocked eager for more information. It's this avid appetite for knowledge and his questioning, which has led to his move from New Mexico State University, to his new role as an Adjunct Professor for the College of Agriculture at Chico State University. At Chico, he will be alongside another formidable soil advocate and artful listener, Dr Tim LaSalle. LaSalle is leading the way in regenerative research and tertiary education.

Johnson's explorations into fungi, began during a USDA project, identifying methods to static compost dairy effluent. They needed a design to help speed up the composting process, use less water and reduce salt in the end product. As a result, a fungal dominant bio-diverse compost (4:1 F:B ratio) was produced. This fungal compost has led Dr. Johnson down an exciting path of discovery, leading him to propose that fungi may be the largest determinant to yields, even more than synthetic nitrogen or phosphorus. His work is showing that with diverse microbial biomass, farming has the potential to produce more than even the most productive natural ecosystems.

He is clear, we have normalised poor soils across the world. "We don't know what highly functional soils can do," he says. "We've been making our observations in crippled soils." These are soils low in carbon with low biodiversity and are bacterially dominated—what we see in nearly all modern chemically farmed or cultivated operations.

He is a proponent for a method he and Hui-Chan have called the Johnson-Su Bioreactor. This Bioreactor involves a simple, slow composting process, which emphasises airflow for optimal fungal growth and worm digestion. Johnson is a man after my own heart, believing "vermicast is really key in this." In his trial plots, a single application of 450kg/ha (400#/ac) of fungal dominated compost, was applied in 2012. Each year, seeds are treated with a liquid compost extract and grown in plots alongside plots planted with conventional NPK fertiliser. His work has measured exponential lifts in yield under biological treatments. For Johnson, "biology is a biomass game," it's not just higher fungal numbers, it's a total increase of diverse organisms (and their quorum signals) in the soil. Johnson's trials involve cover crops, which are either incorporated into the soil or harvested and removed, and the resulting yield gains compared to plots fertilized with NPK have been extraordinary. After 7 years, these plots are drawing in 777 kg/N/Ha year. Other minerals are also increasing calcium (76%), magnesium (84%) zinc (62%) and copper (40%). This mirrors many of the in-field findings with Regenerators discussed in this book.

Bacterially dominated, 'stressed' soils may respire over half of the short-cycle soil carbon back into the atmosphere. Johnson is finding that daily carbon losses to the atmosphere slow, as soils become fungal dominated, with only 11% of C respired. This means the 'good time, not a long time' carbon emissions pool, can be reduced to 20% of current net losses from soil. Regenerative management practices, which foster the growth of beneficial microbes, reduce hardpans, encourage deeper rooting depths and increase plant photosynthesis, are essential in building stable soil carbon. Dr. Johnson's work is measuring up to 10.71 T Mg C ha (-1). This process is the real money in your bank!

With so many variables and movement with carbon, understandably, it gets the scientists all in a twist trying to pin these numbers down. Peter Donovan and Didi Pershouse at the Soil Carbon Coalition are clear, it's time for action, not more debates about whether soil carbon can be drawn down. Sharing data on their open source website, they are helping producers to answer the question--how fast and how efficiently can atmospheric carbon be turned into soil organic matter? We have a substantial opportunity here, through enhancing microbial signaling, fungal biomass and biodiversity. We have the tools to address even the most wicked of issues facing society: climatic variability. What are we waiting for?

In Western Australia, the sun startles everything into brightness. This is the landscape of harsh, high definition contrasts--bright reds, oranges, yellows and startling azure blue skies, punctuated with gumtree greens. When Europeans first settled here 200 years ago, they found "a gentleman's estate" -- a mosaic of grassy savannah, shrublands and tall eucalypts cultivated by over 50,000 years of aboriginal fire management. Many of these native grassland species did not adapt when fire management stopped and set-stocking practices began. The changes in this brittle landscape were rapid. Native species quickly became locally extinct or relegated to remnant outcroppings, roadsides and ditches. The old grandfather trees, 200 years or older, are few and far between and recall a history of a savannah landscape, rather than the dense young eucalypts which now cover unmanaged lands.

On a three-hour trip inland from Perth to visit Bio-Integrity crop farmers Di and Ian Haggerty, at Prospect Farm, you pass the Rabbit Proof Fence into red soil country and the lands of 'silver loams.' The Rabbit Proof Fence was a governmental last gasp (and ultimately futile) attempt at preventing the tide of rabbits from entering the wheatbelt. Built in 1902 it stretched over 1800kms. Anyone who left a gate open, could spend 72 days in prison. Once you pass the Rabbit Proof Fence, the rains stop and the vegetation changes under an unrelenting sun.

There are some who argue Western Australia is not suited to modern ways of farming. With low and variable rainfall, rising salinity, high aluminum, low pH and almost non-existent soil carbon (and don't get me started about those flies), they'd be right. It has been estimated that under each hectare of land in the wheat belt, there lurks a potential 10,000 tonnes of salt. Without tree cover to keep the water table down low, 10% of the wheatbelt is currently too saline to produce food. On the other extreme, up to 24 million ha of land is battling acidic soils with pH values below 4.8. The low pH creates ideal conditions for aluminum toxicity and reduces crop resilience in low rainfall. This low pH and aluminum toxicity is estimated to be costing Western Australia $70 million/yr in lost revenue.

The saying, "farming on the edge" is becoming closer to "farming on a razor blade." From 2000 to 2012, the average annual rainfall across the Prospect Farm blocks averaged 216mm (8.5"). In 2017, vital germination rains arrived too late, so producers were encouraged to chemically spray out their canola and lupin crops, rather than harvest patchy growth. Their agronomists assured them that their fertilisers would still be there next season.

The Haggerty's chose not to spray out, aware that leaving ground bare and vulnerable to wind erosion and weeds didn't make sense. They've never been ones to be pushed into anything that didn't align with nature's cycles. They call their system "Natural Intelligence Farming". They farm 18,000 Ha (44,000 Ac) in the central wheatbelt of Western Australia, with 4 properties within 2 hours drive of each other, which in these Australian landscapes is about the distance locals drive to their letterboxes.

Prospect Farms produces Bio Integrity cereal grains and runs a self-replacing merino flock across their properties. The Haggerty's have developed a farming system that successfully utilises "natural intelligence" via soil microbes, plants, animals and the powerful process of photosynthesis. This system permits them to cost-effectively rehabilitate saline and chemically degraded landscapes. As a side-effect from this process, they're producing high quality grain, meat and fine wool.

Walking into Di and Ian's kitchen, I always have the sense of coming home. The walls of this house don't seem strong enough to contain these minds and personalities, so it seems fit that they are more at home outdoors in the red and silver landscapes. If Di ever tires of farming, there is an occupation for professional hugging waiting for her, she offers these generous embraces to those lucky enough to call her a friend. Ian stands to greet visitors after he unwinds his long-shanked legs from under the kitchen table. His height may seem formidable, until you spend a few moments with this man and see what kind and generous soul dwells inside. This is a house built around good conversations; many have come here and discovered new frontiers in what's possible, when you listen to your land and yourself.

The pair worked out early on in their relationship that having clear lines of responsibility creates a strong foundation for a healthy marriage. As such, Ian is responsible for cropping decisions and Di shepherds their 2,500 quality breeding merino ewes. She has a magic touch with her dogs. Both Ian and Di have a deep feel and connection to their land, crops and animals. The pair have received a lot of press due to their extraordinary results in building soil carbon. Charles Massey calls Di and Ian "revolutionaries" in his 2017 book "The Call of the Reed Warbler," which offers a deep coverage of the growth of Regenerative Agriculture in Australia.

In their youth, both Di and Ian lost the opportunity to continue working on their respective family farms, a point of major stress in their lives. "We had our backs against the wall and there was no hope at all of [returning to] farming." They moved north to the Kimberlys and ran a roadhouse for a few years. During this time, Ian and Di were exposed to novel land management ideas through their friendship with Robyn Tredwell of Birdwood Downs Station. Robyn was the ABC Rural woman of the year in 1995. She advocated using livestock as tools to "Feed, Seed and Weed," along with guidance and ideas from Jane Slattery. These innovative and intuitive ways of interacting with land and livestock, resonated deeply with Ian and Di. The time off the farm also gave them both a good opportunity to reflect and more deeply appreciate the value of managing their own land. In 1994, the chance to purchase 660 Ha in Wyalkatchem, next door to Di's family, was quickly snapped up. The early days created a lot of financial stress, but the opportunity to share-farm with her parents, was too good to pass up and helped them gain a foothold into farming again. When they first arrived, one part of the farm was undergrazed and hadn't been cropped for 30 years. Even though "this part had a lot of saline ground, it was alive and full of mushrooms!" In the first few years, they grew bumper crops, which outshone many in the area.

They quickly learned in this brittle landscape, that neglecting biology, harvesting crops and cultivating, were practices which quickly resulted in declining yields and compaction. The losses in organic matter led to poorer water holding capacities and water repellency; their soils had no capacity to handle dry seasons. Although the field which had not been cropped for 30 years had far better capacity and carbon bank, it too slowly became depleted without any inputs.

The lesson they learned was loud and clear: healthier management was essential to increase moisture retention, carbon and resilience. This message was reinforced by early chance encounters with some of the world's leading regenerative proponents such as Dr. Elaine Ingham, Dr. Christine Jones, Dr. Marten Stapper and American MD and biological agronomist, Dr. Arden Anderson.

The Haggerty's have deep inquisitive minds and a humbleness toward the world around them. It keeps them open to new possibilities, essential skills for regenerative farming success. In the early days, they saw large inconsistencies between their soil and plant tissue tests, sparking a quest to uncover more answers. Their soil tests showed massive mineral deficiencies, yet the plant tests indicated they were at optimal nutrient levels. The plants were able to access what they needed somehow. Their epiphany was triggered early in their farming careers, as they experienced declines in yield, increases in input costs and increasingly unreliable seasons. Nature had this quandary sorted in areas not receiving fertiliser. They took a new focus on soil carbon, building water-holding capacities and looking to how nature's interconnected systems worked. Working with microbes, they realised, is the key to building resilience in a changeable, unpredictable climate.

Their approach focuses on optimising the management of animal grazing and recovery, to maximise the transfer and cycling of nutrients and microbes. Their ground is covered by green growing plants for as much of the year as possible. Little to no synthetic fertilisers, pesticides or fungicides are applied. The pair have never been fond of chemical controls and use their herbicides judiciously. At sowing, the only inputs down the drill are vermiliquids (5 litres/ha or 1/2 gal/ac) and compost extracts (10litres/ha or 1 gal/ac), which optimise the benefits from quorum sensing. The vermiliquid is a commercial product, shipped across from Victoria from the innovative company Nutrisoil. Compost extract is made on site using an imported Mid-West Biosystems extractor and high fungal diverse compost. Having a quality compost or vermicast source is essential to success. This is one area where compromises cannot be made.

They find that the more subtle their approach, the bigger the returns. This sensitivity even extends to the attitude of staff. If people are there for a quick buck and don't value the equipment or the land, their time on the farm will be short. Ian is clear: staff with those attitudes cost the farm in machinery damage, input costs and outcomes in yield and land regeneration slow.

Ian explains his herbicide spray approach to new staff like this: "You drive out into the fields and then you just feel when you need to spray." Feel? That's right, in his massive technologically advanced tractor, Ian is using the most ancient and simplest of approaches: compost extracts and feel. This "feel" is saving the farm in herbicide expenditure and fossil fuel use. As soil structure improves, seed drilling requires less fuel and is easier on the tractor. Ian will fill a spray tank designed to spray 40 hectares and instead cover over 400! They have learned to work with nature's signaling processes, by tickling it with small amounts of extracts.

Prospect Farms has a custodial relationship that values inter-connectedness above and below ground and it shows. In 2010, the Haggerty's took part in an independent CSIRO wheatbelt study looking at soil carbon under cropping. This research showed that the Haggerty farm had significant increases in soil carbon (42%), water holding capacities (13%), nitrogen (28%) and soil minerals, compared to other farms in their area. Remarkably, as their mineral bases lifted, less desired elements, aluminum and sodium, dropped. It took 3 years for the Haggerty data to be included in the main data set. Why? Their data was so different, it was considered an outlier and what do you do to outliers in science? You remove them!!

The farm is drawing down significant amounts of carbon. In the CSIRO study, which was done earlier in their soil health programme they were still drawing an additional 2 tonne/ha/year. Work done by other Australian regenerative farmers, like Colin Seis, shows that carbon drawdown is not a linear relationship, it becomes a reinforcing feedback loop. The Haggerty's expect their carbon levels to be higher now the system has been under their management for longer.

The sunlight capture at Prospect Farms is incredibly efficient, with the Brix levels in wheat running up to 25, outperforming many wheat producers with a plant sap Brix level around 3. Attaining these Brix levels reduces plant frost susceptibility, drawing carbon deep into the soil profile, driving the biological gut system and increasing nutrient density. Frost is becoming an increasing threat to crop production across the planet. The 2018 crops in Western Australia were looking to be a bumper crop. Frost wreaks more havoc on crops in Australia, than fire and hail combined. With many of their neighbours significantly impacted by frost, they only lost 5% across all their blocks. Syntropy here is in clear evidence, life begets more life and increasing resilience.

Prospect Farm is not organic. They still use some herbicides: at half the rates and frequency. All herbicides are buffered with worm extracts, protecting soil microbes against the detrimental effects of the herbicides. Their wheat is quality tested and has no chemical residues, including glyphosate (Round-up), which makes their wheat cleaner than some organic foods on the market. Their grain is tested and contains high levels of true protein, lower nitrates and up to 3 times higher levels of trace elements than conventionally grown grains. The Haggerty's maintain that as their grains have higher hectoliter weight and increased bio-energy levels, this "leads to improved digestibility when our produce is consumed." Consumers around the world are demanding cleaner, nutrient dense foods and the Haggerty's are responding to that demand. Currently, they're providing grain into the Netherlands, with plans to expand production volumes in collaboration with other producers to commence export into South-East Asia.

Switching biological signals on, means new blocks come away naturally, with good inter-row cover of self-seeded sub clovers. Ian has "been really rapt" with the growth of the crops, with zero nitrogen applied. In the 2018 season, neighbouring areas were struggling to access enough N to grow a crop, with some applying around 130 units/Ha (260 pounds urea /ac). At $180/Ha in input costs just in nitrogen, that's not a good start to the season for conventional producers.

Without the need for extensive post-emergent herbicides, pesticides or fungicides, or foliar nutrition, the stress levels have dropped right away for the team. "In this system, we're not running around like everyone else," Ian explains. By leaving more to nature, not worrying about totally clean fields and taking a long-term focus, the stress that used to be a big part of their lives has gone. The downside for Ian now, is having too much time on his hands. This past season he felt the need to do some tractor jockeying. With some nitrogen sitting in the shed, he did a 300-acre foliar application of 6 units of N. Mid-season, he can see it's now the worst performing block on the farm. He's learning to find other things to do with his spare time and leave nature to it!

Shifting how they interact with their land and livestock; means they now leave more than what they're taking out. Challenges that were once seen as curveballs, are now seen as opportunities for learning. This shift in mindset, has changed the trajectory and outcomes for the entire landscape, sending the system into syntropy. Across the entire property, the carbon they were sequestering in 2010 equated to the total global emissions from 23,855 people on the planet every year. This number will have increased this year. They are a striking example of how one family can contribute benefits far beyond the farm gate.

6

Let 'em Breathe

"This land pulses with life. It breathes in me; it breathes around me; it breathes in spite of me. When I walk on this land, I am walking on the heartbeat of the past and the future. And that's only one of the reasons I am a farmer."
– Brenda Sutton Rose

If you ask most fertiliser salesmen what the number one yield limiting factor is, they'll tell you it's nitrogen. Ask a farmer or rancher and they'll say it's water. Well, try this... hold your hand tightly over your nose and mouth for a few minutes to see what your number one factor to survival is... It's AIR.

Just as you won't thrive for long without air, this is also true for most plants and many beneficial soil microbes. Along with forests, soil functions like the planet's lungs, drawing air in and out in a daily rhythm. Without adequate airflow, roots and microbes curl up and die, and vital mineral and water cycles breakdown. Compacted and waterlogged soils tie-up or lose valuable nutrients including nitrogen. Have you ever noticed how grass grows shorter in vehicle tracks in a field or along well-trodden paths? Many farms and ranches can have this effect over their whole place, masking the effect of compaction.

In the triage of identifying critical actions we can take, the first step in improving plant performance starts with soil that can breathe. When soils are not well aerated, then water and fertiliser inputs must be increased. Air (and water), moves into soil through the gaps in soil aggregates; the crumbs formed by microbes. Consider a healthy soil structure like the construction of a city block. Termites, dung beetles, ants and earthworms build the major structures, the city streets and gaps between buildings. The micro-critters, the protists and nematodes, make the hallways, stairwells and the living spaces between the crumbs. The bacteria and fungi build the mortar and bricks to construct the walls and floors. This soil structure also includes essential places; imagine banks, schools, hospitals and even a pub (saloon). Imagine what happens when a community loses their services, especially the pub! Communities rapidly decline without these essential services. Through soil compaction, critical services for microbes and plants degrade, creating a vicious downwards cycle. Poor soil structure turns these apartments into a tarmac. This poor airflow and loss of structure stalls natural cycles in the soil, water, carbon, nitrogen, phosphorus, sulfur, etc. You do not want a soil that cannot breathe freely!

Most properties I visit have major issues with airflow. Indeed, digging holes on New Zealand dairy farms reveals a deep cause for alarm. A recent government report showed that nearly 80% of all dairy farms were badly affected by compaction.[21] This is a concern, as compacted soils require more water and more nitrogen. All of this means, these farms are requiring increasing inputs to maintain production. And just to really hit another nail into the dairy coffin, these soils may be losing 10 times as much N into the atmosphere and waterways. Hence, some dairy farms in NZ are spending upwards of $900 NZD/ha ($250USD/ac) to grow a blade of grass.

Compaction can create a boom/bust cycle as bacteria run out of food. Upon their death they release nitrite. Not all forms of nitrogen are the same. Nitrites and nitrates move quickly through soil to be released to waterways or the atmosphere. In excess, they create poor quality forage and in worse case scenarios, can result in animal and even human deaths. Nitrite/nitrates stimulate the germination of many weeds, such as cape daisy and low-quality grasses such as Rat's-tail and Barley grass. Protists are feeding on fungi and bacteria, so a programme which builds bacterial numbers will build protists. 75% of plant available N comes from protists grazing on bacteria.

How can you tell if your soils are compacted? Digging a hole is a good start. Compacted soils often have plate-like structures, roots will be shallow, shear off or travel sideways. Other indicators include fine soil crumbs, surface crusting, thatch and high insect or disease pressures. These areas will have higher water stress, less plant growth and slower recovery. In compacted areas, water will pond or run off, plants such as moss or deep-rooted weeds will grow.

Want to get a clearer more accurate measure for compaction? A penetrometer is a tool you can purchase on-line and use to monitor current conditions and measure changes over time. It gives you a reading based on how many PSI it takes for a root to push through the soil. Take your penetrometer and push it into soil. Ideally, do this when your soil is at a Goldilocks moisture level, not too wet and not too dry (generally spring following rains or snow melt). Don't push too fast, ideally one inch per second.

As resistance increases beyond 300PSI, root penetration drops. There may still be some roots that will get through that layer, but they're generally sparse and in poor shape. Being able to open up soils and increase root penetrations like this, has huge broad-reaching positive implications well beyond the farm gate, improving water quality and reducing greenhouse gases.

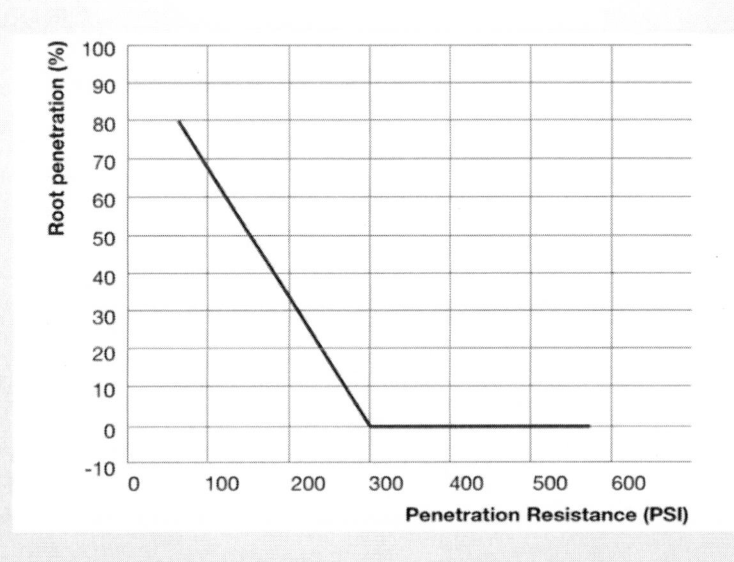

Picture two apple orchards side by side; separated only by a headland and a gulf in paradigms. Both orchards were planted at the same time. One is managed by a New Zealand research station investigating best management practices and integrated pest management (IPM); the other had been in organic production for 7 years. The IPM orchard is irrigated, grass between rows is maintained like a golf turf, with a nice tidy herbicide strip under the trees. Meanwhile, the organic orchard is an unruly explosion of grass and colour. Every year, 5-10 m3/Ha compost has been spread under the trees and there is no irrigation here; it's just not needed.

Organic orchardist Nick Pattison is a laconic character, belying a deep passionate concern for people and planet. We first met at a Dr Arden Anderson soil school. I arrived solo, armed only with a university degree in soils and 5 years managing gardens and orchards. I left the school with the profound realisation that my understanding of the underground world amounted to squat. Every evening of the school involved a rowdy round of drinks, with what became my tight crew of influencers and lifelong friends, Steve Erickson, Tom Harris and Nick. Nick had been battling demons for years when we first met. His teenage daughter had been killed in a tragic car accident one rainy night. I think our friendship drew Nick back from the darkness. As a pioneer in organic pipfruit production, he had accumulated years of learning, which he would share over bottles of wine and animated cigarette jabs.

Before shifting to organic production in 1997, Nick's parents were also living on the property, enabling Nick to care for his elderly mother during her final stages of terminal cancer. One Spring day, as Nick sprayed a control for mealy bug on his Black Doris plums, he looked up to see his mum hanging out washing; as the chemicals slowly drifted around her.

The insecticide he was using was Tokuthion, an organophosphate. These chemicals have been linked to cancers, chronic fatigue syndrome, Parkinson's disease and neurological impairment in the womb and in young babies.[22] The mealy bugs had been a thorn in Nick's side for years and frustratingly the powerful insecticide was not keeping them at bay.

Nick was aware of some potential dangers from these chemicals. However, seeing his mother haloed by the chemicals, provided the incentive he needed to drop all organophosphates that year. The following year, the mealy bug that had been causing significant crop damage dropped in numbers and the next year... they were totally gone. This process is known as Trophobiosis, whereby chemical inputs disrupt plant physiology, leading them to be more susceptible to pests and disease. This insight threw Nick's entire world view on the ground, wiping his hands of any soluble chemical interventions. Within a few years, the orchard was fully certified organic. In the early days, he adopted what he calls a "light green" approach to organic production, substituting harsh chemicals for their organic counterparts.

Nick was one of a small group of pioneering Pipfruit producers, who were motivated to convert to certified Organic production in 1997. He reflects, "it was pretty brave, looking back." One driver for change, was the desire to break the Apple and Pear Board monopoly on the sector, plus Organic Pipfruit was offering a sizeable premium at the time.

In the 1990s, Pipfruit producers were controlled by the Apple and Pear Board. The board specified what types of chemicals and application rates and timings; if a grower wanted to sell their fruit, they had to strictly adhere to this spray calendar. "There was no monitoring or oversight and no regard to whether sprays were needed or not." This calendar put growers on "an automaton setting, it totally disempowered us." Nick felt that the use of organophosphates was excessive, "it was like taking a sledgehammer to a thumbtack."

Looking for organic nitrogen inputs, Nick turned his eye and machinery, to compost production. He attended an "inspiring" workshop with American microbiologist Dr. Elaine Ingham, Director and mastermind behind the Soil Foodweb Laboratory. Nick found the science and knowledge he was looking for, giving him the tools and know-how required to produce beautiful top quality high-fungal compost. Arriving to Nick's compost site was one of my favourite downtime activities, with the rich smells that always remind me of rolling bread dough and walking through old growth forests. Making compost is a grown-man's childhood fantasy, playing in a giant sandpit with an articulated caterpillar loader, tractors and compost turner. Ok, maybe not just grown men, I could spend days contentedly building, rolling and moving piles of compost. Through using compost on his own orchards, he discovered that compost was not just an organic mulch or nitrogen source, it was a rich inoculant for a diverse array of beneficial microbes.

One of the organisms you'll often see in compost, appears like a white ash layer. Look closely, however and you'll see it resembles fine hyphal threads. These belong to a group of long-chain bacteria, called actinobacteria. You've possibly had an intimate relationship with these bacteria, as they produce 80% of the different types of medical antibiotics;[23] take streptomyces, they make streptomycin. It's these organisms that contribute to the distinct smell of soil and compost, called 'geosmin.' Geosmin literally translates as the "odour of the earth." When geosmin is combined with the other molecules emitted from plants and rocks, the scent is called 'petrichor.' A word taken from the Greek 'petra' for "stone" and 'ichor' the substance flowing in the veins of the gods. It's what gives beetroot[24] its taste and what creates the smell after a shower of rain meets the road on a hot Summer's day. You're smelling the actinobacteria celebrate, by throwing their spores into the air. Even the most nose-blind city dweller is tuned to sense geosmin, at 5 parts per trillion. To put this into context, I'll let Dr Hank Campbell, President of the American Council on Science and Health, describe it to you; "a shark can smell blood at one part per million. That means human noses are 200,000X more sensitive to geosmin...than a shark is to blood."[25] Why are we so geared to smell it, scientists have no idea. Perhaps we can smell geosmin, as we know it's so good for us; actinobacteria have been linked to reductions in stress and PTSD in humans.[26] Or maybe we can smell it so vividly, as it's there to remind us about where we all came from. We are, after all, more bacterial in our DNA, than human.

Another important bacterium involved in decomposing organic matter, *Mycobacterium vaccae,* is being proposed as a potential PTSD vaccine for first responders and active military. It's astounding to consider, that these examples represent just a few stars in a galaxy of potentially billions of soil organisms that we are yet to comprehend. It is possible, these hidden treasures hold the keys for the biggest challenges which face humanity. I believe while Nick was making some of the best biological inoculants, he was also healing his heart. In a reflective twist, he later discarded the drinking and cigarettes, in favour of the rush of endorphins provided by his new Yoga habit.

For years, Nick provided composting and compost tea services to other horticulturalists, while applying 20 T of compost to his own orchard. Besides compost, the orchard also received lime, every 3 years at 400kgs/Ha (350lbs/Ac) and a fish/humic blend with added beneficial microbiology created by my other inspirer, the renegade fish fertiliser genius, Tom Harris in Nelson, NZ. [6]

---

[6] Sadly, Tom Harris passed away in February 2017, taking with him an unmatched depth of knowledge around soil microbial systems, soil and plant health. He is sorely missed. His passing was a loss to the entire world.

During the 2006 season, a Government Research facility, Plant and Food, began a study into how soil structure differed under different management practices. Nick's place was the ideal choice; it had the same soil type, texture and previous land-use history and even better, it was right next door. The researchers, Marcus Deurer, Karin Müller and Brent Clothier were pioneering an advanced (and exciting!) piece of technology; a 3D X-ray computed tomography. Basically, it's a fancy piece of equipment which takes x-ray images of the empty pore spaces in soil.[27]

The soil carbon measurements were substantially different. Nick's Orchard had 32% more soil organic carbon (SOC). However, what was even more attention-grabbing was what they saw with the X-ray machine. Typically, when you learn about soils at school or university, you'll see soil makeup expressed like this; 45% minerals, 25% air, 25% water and 5% organic matter (OM). However, the pore spaces in Nick's soil were 78%, well over the 50% air and water we're taught in soils school.

As the x-ray images on the following page show, his soils have more air than soil! The researchers found striking results, apart from better aeration to soil and roots, these soils produced less and emitted less nitrous oxide, a green-house gas, and had a 12 tonne/Ha increase in soil carbon in the top 100mm (even though there is far less volume of soil!). The organic orchard compared to the IPM, had a functional water cycle, hence no need for irrigation. These pore spaces store water and function like a sponge, rapidly drinking water into the soil. After heavy rainfall, events Nick would see large ponds of water sitting on the neighbouring block while his soil stayed spongy; he was able to drive equipment on it without damage to soil. These pores do more than just let in water, they are full of biology which also help to filter contaminants and mix nutrients.

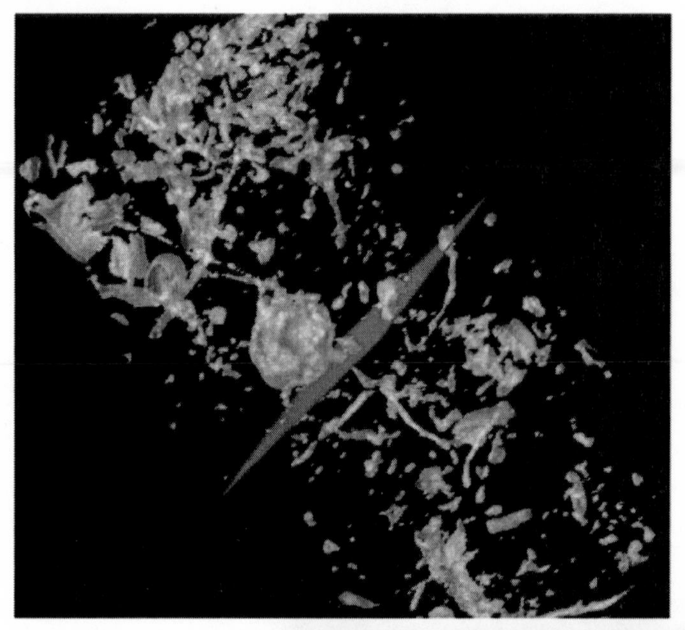

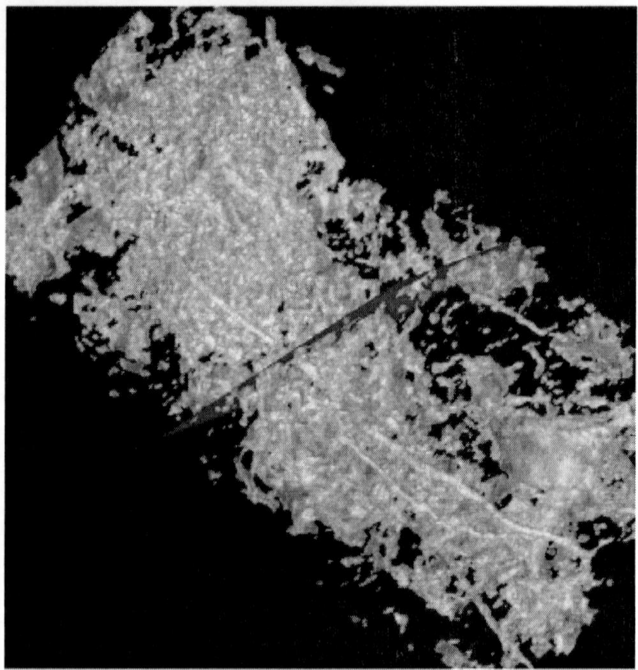

Image: *3D X-ray computed tomography of the two soils. Soil at the top is from the IPM, on bottom is Nick Patterson's. Image with Permission from Plant and Food Research New Zealand.*

In the two different soils, researchers also measured the fate of chemicals, so they applied 2,4-D to see what would happen. 2,4-D is a residual synthetic plant hormone used to control broadleaf weeds; found in over 1,500 commonly used herbicides today (it was also one of the agents in agent orange). There are links between exposure to 2,4-D and sarcoma (a soft-tissue cancer) and non-Hodgkin's lymphoma, making it a suspected carcinogen, however the jury is still out on this one. 2,4-D has a half-life of 280 years. [28] This residual nature is creating long lasting impacts in the environment. It is one of many chemicals contributing to the decline of the Great Barrier Reef, another one of those 'wicked' problems. As reefs are made of a symbiotic relationship between a simple plants and animals (coral polyps)[29] runoff containing herbicides is causing real concern.

On the IPM orchard, the 2,4-D stayed active on the soil surface and with a rainfall event would wash into drains before flowing to the ocean. In the organic orchard, they measured the soil after application and found no trace of the chemical, so they applied it again, with the same result. Where had the chemical gone? It had been bioremediated. This means the soil microbes had consumed the 2,4-D and broken it down into component parts, which then no longer present a threat to the environment. It is essential that we have soils that function like this!

Nick continued to grow apples for a few more years after this study. Through monitoring and management, he successfully cut lime, sulphur and fungicides in half and never applied copper. However, he struggled with the same issues many innovative producers battle; using organic practices for crop varieties, which had been bred for flavour and bred to produce high volumes of fruit under chemical regimes. This breeding had selected against trees natural defense to pests and diseases. The options for many organic producers were limited; pull the trees out and replant, continue to apply multiple lime sulphur sprays with declines in plant (and soil health), change crops, or sell out to the corporate fruit growers. In 2008-2011 seasons, with a perfect storm of restrictions in market access, dropping premiums and massive crop losses to frost and wind, only 3 family farm growers were left in the region, replaced by the mighty corporates. Nick chose lifestyle over the demands of orcharding, pulling the apple trees out and replacing them with an array of specimen trees and sculpted landscaping. The farm now offers niche accommodation in their rural sanctuary and a fascinating contribution to soil science and regenerative agriculture has ended.

One measure for regenerative success is an increase in topsoil, the dark chocolaty brown top layer in soils. The topsoil you start with, is going to depend on historic and present land use, climate, parent materials and historic vegetation. The deepest most beautiful and productive topsoils are found in regions with long geological grassland histories and good seasonal rainfall; what are now the lands of corn and soy in the US, veldts of South Africa, the downs of Australia and the pampas of Argentina and Uruguay. There are also deep soils under the steppes of the former Soviet Union, parts of Northern US, Canada and China. The term 'steppes' is used to describe the short grass prairies, found in lower rainfall, colder climates. The deepest soils I've ever had the pleasure to dig in, were at the 9,000 ha (22,000 ac) Northfork Ranch, alongside Glacier National Park near the Canadian border. The land is surrounded by the 1.5 million-acre Blackfeet Reservation. These First Nation people, were so named either for the black colour of their moccasins, darkened with ashes, or because their main food resource, the Bison, had black feet. From the west, the area is reliably supplied by pristine glacial waters from the Rockies. With the warm chinook winds periodically sending an arc across the skies, the climate had long supported a high protein fescue grass and one of the richest buffalo biomes in North America. Ranchers and farmers all over the world will tell you how challenging it is to grow food in their individual region. Now this ranch is a demonstration of the challenges people and nature can dish out, challenges which range from particular neighbours who cut fences to graze their horses on stockpiled feed, to grazing pressure from thousands of elk, 30-foot snowbanks and temperatures that can drop from 10 to 20 °C (50–68 °F) to below–20 °C (–4 °F) in just a few hours. Throw into the mix the predators--wolves, hawks, coyotes--and then every Thursday night during Spring...takeout night for a resident grizzly bear with a highly developed taste for veal. Now those are challenging growing conditions!

With the help of local Tribal and Federal fish wildlife and parks authorities, the ranchers had been setting bear traps for months; impressive contraptions, with huge spring steel doors and large signs reading, "Caution. Do not enter." Sadly, I think their bears could read and the traps remained empty.

In the centre of a calving field, their resident bear had staked his claim in a den 20 feet down an embankment. Entering the cave (after the rancher had declared it safe!), I discovered the cave was nestled into a deep loamy brown soil. "How long does it take to form a soil like that?" This question has as many answers as there are soils. Most soil scientists agree that it takes between 100-500 years to grow an inch of topsoil, depending on vegetation, climate and other soil-forming factors. If you have a wet, hot climate like Argentina, then soils will form relatively quickly compared to cold, arid environments like the high deserts in Nevada.

On the southern headland of Auckland's Manukau harbor, are some of New Zealand's most productive and most expensive, rural lands. Deep volcanic soils nurtured by the good rainfall, once supported the highest density of prime Māori gardens for 5 centuries prior to European arrival. On a rare windless and sunny day, the view truly is a postcard picture of water and forests. You can almost forget how close you are to the noise and traffic jams of the city. In 2015, we went on a troubleshooting mission at Lindsay Farm with land managers who have caught the soil bug.

The farm originally was a dry stock (finishing) farm, before being developed into a top-of-the-line horse breeding, training and racing facility. The farm covers over 220 Ha (550 acres), which also runs sheep and cattle for pasture management. When laying the foundation for their valuable racehorses, extensive earthworks were done on the rolling hills. The earthmovers filled gullies and smoothed off troublesome hilltops, which on the surface looks impressive. Dig a little deeper, however and it becomes apparent, all is not well. Earthworks interrupted groundwater flows and caused extensive compaction and soil structure damage. Underground erosion created cracks, posing a risk to racehorses prone to frolicking around. In higher areas, topsoil was razed off, exposing heavy clays, to which only a scattering of topsoil was applied before Ryegrass was reseeded. James White, the current farm manager, felt the earthworks "ripped the heart out of this place." Soil mineral and microbial imbalances and compaction zones created ideal conditions for insect pest invaders. Not only were soils eroding due to the mechanical disruptions, soil cracks were also created by mineral imbalances.

Chemically, the historic soil tests looked to be in fairly good shape. Except for one glaring factor, to keep the place picture-perfect, they were irrigating during Summer, unaware that the costly pod irrigation system was drawing water from a saline bore. As James pointed out, salty water and heavy clay are not a good mix. On our visit, the first signs of the sodium were just starting to show, with the ornamental trees lining the entrance ways showing die-back and burn on the leaf edges. The combination of mineral and microbial imbalances and physical constraints was creating detrimental effects on soil function, pasture health and animal health.

Around Auckland, I'm used to seeing compaction, however this property was taking it to a whole new level. Close to the main barn, grasses were low growing and shallow rooted, with flat weeds prevailing. With the poor air movement, biological diversity and soil moisture holding capacities were low. This creates soil conditions like many feedlot operations experience; from Winter mud to Summer concrete and little transition time between. Compacted soils also leave water with little option but to run off. With little or no vegetation cover, erosion becomes a vicious cycle.

With the best intentions in the world, these pastures were far from what was needed to produce Group 1 winners. The bloodstock manager was seeing some behavioral issues and ticks were an increasing problem, horse coats looked rough despite some of the best supplements on the market.

During the site visit, plant tissue, soil mineral and microbial testing identified many opportunities for action. The areas we aimed to target, were the poor air movement, poor infiltration, thatching, short root systems, grass pull, low Brix, insect pests and plant diseases. The timing to take on a soil health programme was perfect!

In our triage, AIR flow had been critically disrupted. Good soil structure is the result of interactions between chemical, biological and physical factors, all underpinned by management. Management practices that can increase compaction include overgrazing, chemical use, poor irrigation practices, cultivation, lack of species diversity, chemical/bare fallow, heavy machinery in wet soils etc.

With any soil health programme, we always begin by rolling out methods which either avoid or repair compaction events. Dig and discover this for yourself--is compaction shallow? Is it a "cow pan" in the top 5 cm (2 inches) from animal impact (like horse hooves!) Or is it a deep hardpan caused by machinery, soil types or fine soil particle accumulation?

On Lindsay Farm, there were both types of compaction. Shallow compaction and surface crusting, can be addressed from the top-down through minerals, lifting plant Brix and encouraging critters to build the apartment. Cover crops are an excellent tool to reduce compaction issues. On this farm however, cover crops were not a tool available to us; they do not fit the aesthetic paradigm for a racehorse property.

We had a rare opportunity to fast-track changes to make significant inroads to achieve our soil health goals. So, we hit all spheres of the soil structure dynamic. Through 1) aeration, 2) balancing the soil chemical profile, 3) addressing insect pests, 4) building carbon and Brix and 5) removing or buffering harmful inputs.

1) Aeration

Given the severe compaction and shallow topsoil following earthworks, we chose mechanical intervention. This is not a choice I take lightly, but it can be done with biological processes in mind, as a *once-off* remediation. It is important that any mechanical intervention is carried out when the soil has optimum moisture for the task in hand, otherwise efforts can end up being detrimental. As humans, we love the big shiny equipment, so there is a glut of mechanical options to choose from. Equipment choices need to consider your specific conditions. At Lindsay Farm, pasture was already established, so we needed to ensure grass had a minimum amount of disturbance. Sub-soiling in the spring would run the risk of creating more cracks and other associated problems.

For this situation, we recommended two mechanical methods to improve soil structure. First, breaking up the soil crust with harrows and secondly, by aerating the top 150mm (6in) to open-up subsoil. This will immediately improve drainage and root penetration. Initially, harrows were used to scuff up thatch and any dead plant materials, breaking up the soil surface to aid infiltration.

Anytime, in fact *EVERY* time, we disturb soil, microbes wake up hungry. If you don't feed them, bacterial populations boom, as they feed upon the exposed soil carbon. Generally, any mechanical rips can seal up quickly again, especially if the underlying conditions which led to compaction are not addressed. We always 'rip and drip!' You can feed your microbes with something simple and cheap, like molasses and humic acid at low rates (I'm talking a few cups/Ha (one cup/Ac) and a litre of humic). Investigate how you can modify equipment to drip liquids at low rates down the tines. I've seen producers mount 200 litre drums (50 gal) on top of aerators or carry them in the tractor bucket. Into the drum, you can insert hoses with taps to ensure it's just a slow trickle of liquid down the rip lines. At Lindsay Farm, fish, seaweed and a liquid lime was dripped to help keep the pathways open. Their root systems can now penetrate and get on with their job; building soil.

If you're not limited to having lush lawn-like pastures, then longer term solutions to tight compacted soils, may include the introduction of diverse deep tap-rooted species like Lucerne (alfalfa), sunflowers and chicory. Many annual grasses, such as oats and rye, have fibrous roots to open-up tight soil and can be used as a first step before putting in perennial pastures.

2) Mineral imbalances

Achieving soil health goals is a 'Catch 22.' When soils are compacted, biological action is inhibited and the critters who build soil structure are stopped in their tracks. With surface crusting, shallow roots, limited topsoil and the high sodium at Lindsay Farm, we chose to apply solid rates of gypsum between 500-1000 kg/ha (450-900 #/ac) in the first year. Gypsum ($CaSO_4$) is a powerful tool when looking to drop excess cations and to flocculate or open-up soils. I think of gypsum a little like flushing a toilet (beginning to wonder if I have an ablutions obsession!) When gypsum is applied, it displaces the sodium ion and replaces it with calcium. We kept an eye on magnesium as well, as the gypsum has the same effect on Mg. It's a little tricky using gypsum on soils which have an impervious layer, as you need the sodium to be flushed somewhere else, back to the sea in the Lindsay Farm case! To mitigate the high salt levels from the bore water, a humate was also added to the gypsum, to buffer sodium's effect on the plant. When in doubt, use carbon!

During March and April 2017, Auckland City experienced unprecedented rainfall events, which reduced the need for irrigation. Combined with the soil inputs, sodium levels rapidly dropped to the levels they were at before brine was applied. The graph here shows the influence of the irrigation which started after the 2011 soil tests were taken. After the 2015 tests were taken, we began our soil rehabilitation programme.

Taking tests like this (and having someone to work with you to interpret the results), effectively meant a crisis was nipped in the bud.

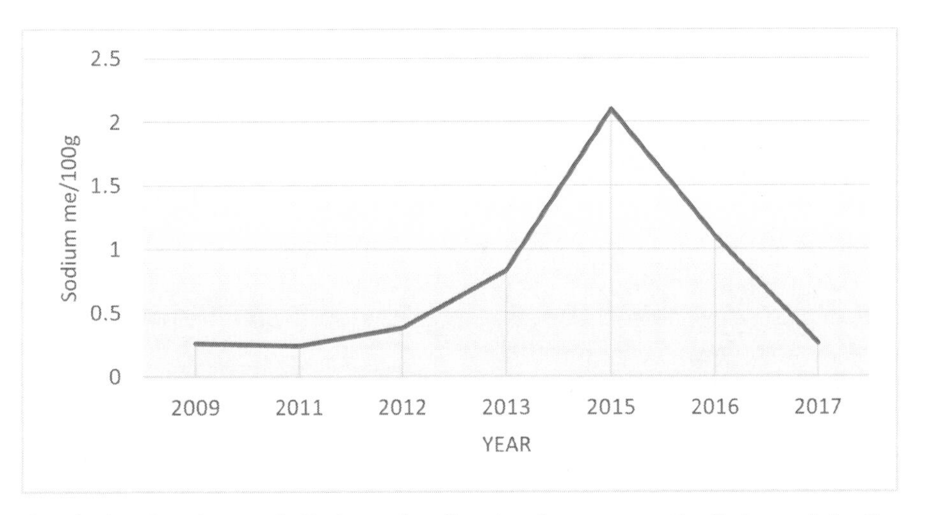

Graph showing changes in Spring soil sodium levels. 2012 was the first-year irrigating with brine, late 2015 biological products applied.

3) Building Carbon

During site visits with a coach from Integrity Soils, Phil Billings, we spotted large volumes of compostable materials which were being hauled offsite or dumped on-farm and wasted. Every week the farm was generating between 11 to 15 m3 of waste including sawdust, manure, lawn clippings and tree waste. Over a year, this added up to over 780 m^3 (yd^3). We humans love to create waste problems; however, in nature, there is no such thing as waste, just opportunities. We found the perfect site for on-farm composting and within six months the managers were creating a top-quality compost. Every year, this waste had been costing the owner tens of thousands of dollars to dispose. This saving easily paid for composting equipment, labour and applications. Most areas on the main farm receive 10 m^3/ha (4 yd^3/ac) of compost annually.

As we explored earlier; the plant sending sugars out into the rhizosphere, is what builds stable soil carbon deeper into the profile. Plants need to be healthy and photosynthesising optimally, or they will not allocate valuable sugars to exudates. The plant Brix at Lindsay Farm were all low, between 4 and 5, before the soil programme began, doubling by year two. This was achieved by addressing compaction, feeding the soil/plant with fish and humic and knocking out pests to give plants a chance.

4) Addressing insect pests

The property was infested with leaf- and root-chewing or sucking insects. They had more livestock eating the pasture belowground, than aboveground! Steps were taken to knock these insects back using bio-controls, and then working to address the underlying soil factors which were attracting insects: namely, the low Brix, incomplete proteins, and compaction. We will delve into causal processes and actions for pests and insects in the following chapters.

5) Just cut it out

Many popular fertilisers and 'i-cides negatively impact plant physiology, soil structure and/or microbial communities. What gave Organics a bad name in rural communities, was the tendency to pull the rug out on inputs, before soil, plant and animal health were functioning optimally. As the biological agronomist Gary Zimmer says, "You need to earn the right" to remove the chemical props. Becoming certified Organic, is not just a question of turning off the chemicals. Just as a meat eater does not become vegetarian overnight, by just omitting meat from their diet (well actually you could, but there will be health consequences), this switching-off approach, meant many organic producers dropped inputs before they'd earned the right. As a result, yields dropped, weeds grew, and animal health was compromised. There is no need to go cold turkey; there are perfectly good alternatives or options to help you mitigate harm during the transition phase. At Lindsay Farm we removed the soluble MAP (Monoammonium phosphate) which suppress mycorrhizae, exchanging it with bio-friendly phosphate, in guano and fish hydrolysate. The urea was replaced by fish and a small amount of calcium nitrate, and herbicides were buffered using fulvic acid. What does buffering mean? In these terms, I mean something which smooths impact or reduces shock to microbes, soil and plants. By using a fulvic acid with chemicals, there is a multi-pronged beneficial effect.

Fulvic and humic acids increase cell wall permeability by 30%.[30] Plants recognise the carbon and food benefits and open themselves up, yummy – resulting in a more effective efficient kill, or nutrient uptake, when using foliar nutrition. Fulvic acid enables you to reduce the rate of most chemicals by 30%, with no reduction in efficacy. Fulvic acid also stimulates the breakdown of residual herbicides via photo-oxidation.[31] In broadacre cropping, this technique is enabling producers to use minimal herbicide applications and still grow nil-residue crops. With attention to a soil health programme, herbicide resistance is not a concern.

Commercial humic and fulvic acid products are commonly extracted from non-renewable soft coal seams. Renewable humate forms can be extracted from compost or vermicast (worm compost). We regularly use humates in soil and animal (and human) health programs. Humic acids provide wide and varied benefits for plant health; increasing the communication pathways involved in plant development (prehistoric Quorum Signals), leading to increased root growth, yield, health and quality of the foods produced.

Generally, an approach to rehabilitating degraded soils takes commitment, patience and a pinch of faith. Changes commonly happen below-ground first, often well before above ground signs. This has been a communication failing for me over the years. Producers will set up a strip in the middle of the field to trial and then after one year, give up, saying "visually there was no difference." It may have taken decades, if not hundreds of years, for soils to become degraded and it can be a little challenging working with producers, who expect to see miracles in their first year. Creating realistic expectations is important with producers accustomed to the instant results from chemicals.

Fortunately, this was not the case at Lindsay Farm. Our early monitoring of the soil health programme far exceeded everyone's initial expectations, which is always a bonus! Remember how I said scientists say it can take between 100-500 years to grow an inch of soil? In areas where topsoil had been removed, we measured 18-22 mm (0.7-0.86") increase in topsoil depth in an 18-month window. That's 1mm every month!

These below-ground changes resulted in visibly remarkable shifts in water-holding capacity, nutrient availability/ cycling and pasture density. The insect controls were hugely successful, leading to more clover, more grass and deeper root penetration. These factors resulted in a reduced need for irrigation. James says the improvements are also being reflected in pasture quality and horse health. "Our bloodstock agent made a comment that the weanlings are looking the best he's seen them. We can only put that down to the grass that we're growing from healthier soil."

Other livestock have also leaped ahead, with Lindsay Farm receiving Integrity Soils' first 'Worm of the Week Award'. "Two years ago," James reflects, "we had very little worm activity and a lot of insect pressures. From memory, on a one-foot square and one-foot deep count, we had five earthworms. [This time], we had over 70!" The award sits with pride alongside cabinets jammed full of racing trophies and accolades. It's James's personal favourite!

The changes at Lindsay Farm, are a real credit to the work and commitment of the farm management team. Through timely use of aeration, compost applications, grazing management and applying inputs as designed, they fast-tracked their success. The team invested into the 5 M's, building soils using multiple tools to optimise microbial, mineral, OM, management and yes, even mindset! My mind boggles at the possibilities, if all farms and ranches had passionate focused stewards like this.

Update: our close relationship with farm staff hit an unexpected hurdle, when internal personal management issues arose. As a result, our professional relationship has been put on hold. With the investment into capital fertiliser and addressing compaction, we expect the results will continue on the same trajectory with different management.

In conventional production systems, if crop growth is poor, salesmen will generally reach to nitrogen to quick fix the problem. At 78%, nitrogen is the most common element in the atmosphere. It often strikes me as bizarre, that we're applying the amount of fertiliser that we are. In a well-aerated, biologically diverse soil, nitrogen is not a major limit to production. The most productive ecosystems, with the highest biomass production (picture the native tall grass prairies and rainforests), do not require any nitrogen fertilisers.

Before World War One, the Germans were concerned about access to rock reserves of potassium nitrate required for bomb production. The development of the Haber Bosch process solved this issue, breaking the triple bond of atmospheric Nitrogen (N^2) under intense heat and pressure to produce synthetic ammonia. This process occurs efficiently in nature; for humans however, the method is incredibly energy intensive. Chemical nitrogen processes use around 4% of the world's entire natural gas reserves. You know the rest of the story, at the end of the war, the application of synthetic N provided the fuel to explode the human population. Around half of the nitrogen in your body, probably originated from the Haber Bosch [32] process, unless you're eating organic food. Millions have died from Fritz Haber's explosives, and nobly, billions have been fed. However, Haber did not foresee cascading "unintended consequences," with greenhouse gas emissions, water quality, soil erosion, loss in resilience and declining food quality, as a direct result.

There is no argument, soluble N does grow plants, and globally we're been using more of it every year. Unfortunately, this approach is also incredibly leaky and inefficient. Additionally, as there are no checks and balances for N use, someone else is picking up the tab further downstream (or in the air). Despite raising awareness, legislative threats and directives, N-use has continued to escalate over the past 4 decades. The only time consumption dropped in New Zealand? In a year when nitrogen prices escalated.

I've talked about how soil systems are like gut systems and many have constipation or gas. In NZ and other high production high rainfall environments, the soils have something akin to diarrhea. Many high-intensity properties are losing more than 80 kg of N/Ha/yr.[33] This may be conservative, as some NZ dairy farms are applying over 350kg of actual N/Ha every year! As the average dairy farm only uses 15-35% of every unit applied, the majority of N is being lost to the air and waterways (globally this figure is

5-15%).[34] Just pause and consider that for a moment. 5-15%, that's an insane figure. There wouldn't be many businesses happy with those kinds of inefficiencies, particularly for something which may be such a major input. So why do we tolerate it in farming?

In response to environmental concerns, fertiliser companies are focusing their efforts on the "plug-in" products to improve N efficiencies, like DCD, Nitrapyrin and Agrotain. Even with the best projections, using best practices and the best farmers, the soundest estimates, offer a 60% efficiency, with more unintended consequences. DCD was the N-inhibitor I mentioned earlier, which was pulled when it was discovered in milk and in rivers. These plug-ins enable fertiliser companies to continue business as usual, without addressing the key issue; why do you need to add soluble N, and why is your nitrogen cycle not working?

The high use of soluble nitrogen, creates an entropic system, a chaotic vicious cycle, with decreasing returns, due to the breakdown of soil carbon and humus. Microbial communities are altered with a decline in beneficial fungi and the free-living N-fixers, with an increase in the bacteria that love to feed upon N. This loss of carbon creates the conditions for compaction, increasing runoff and erosion and limiting root growth. Just to really put the boot in, these soils then require more irrigation, increasing the vulnerability of farm systems.[35]

Nitrogen applications reduce Brix and the amount of carbon being sent out the roots to feed our beneficial workforce.[36] Once while visiting a research facility, I mentioned how Brix drops with nitrogen; the skeptical scientists scoffed. I invited them to harvest 5 x 3-inch samples from a trial area which had received different rates of N, from the control of 0 to 400 kgs. Without knowledge of which plants had come from which area, I laid them out in order of nitrogen rates; the highest Brix was from the control and with each increase in N rate, the Brix dropped. I don't know what the researchers thought, but it was affirming for me.

Often when considering natural nitrogen inputs, most producers think of legumes, particularly clover or manures and the bacterial genus rhizobia for N fixation. However, in healthy soils the diazotrophs, free-living bacteria which fix nitrogen into the soil are common, such as Azotobacter and Frankia species. These free-living N fixers require a soil with a functional gas exchange system. When soils are compacted or are fine and structureless, N-fixation is compromised.

When plant growth is slow and leaves are yellow, producers often assume that it's a nitrogen deficiency, however it may also be iron (in younger leaves), potassium, magnesium, molybdenum, sulphur, manganese or zinc, or other environmental factors. Don't just rely on plant clues for N, dig a hole and look for the lovely chocolate cake crumble. This crumbly structure should stick to plant roots to form the Rastafarian root systems. These structures provide a useful tool to assess whether the nitrogen cycle is working, or not. Many essential free-living bacteria live inside the soil crumbs. Here, the environment is similar to the nodule of a root. These structures are vital for N-fixation, and any loss in aggregation, or compaction, shuts this process down. Many properties have lost this essential soil aggregation; (re)building soil structure becomes our initial triage.

The path to successfully regenerating soils, is through enhancing natural cycles and using proactive practices which address root causes, not the symptoms in a degraded system. The Regenerators foster their underground livestock, enabling them to profitably reduce nitrogen inputs. A key tool in being able to reduce N, is through the addition of carbon-based foods. Humic substances buffer and chelate N, reducing atmospheric and leaching losses and also increase nitrogen fixation by native N fixers.[37] Bio-stimulants improve the N-cycle through improving soil structure and nitrogen storage,[38] whilst maintaining yields. [39] [40]

Anytime (actually *every* time) you use nitrogen, add some carbon. The two go hand-in-hand, in specific ratios, in every living organism on the planet: from bacteria, with a narrow ratio between 2-7:1 C:N, fungi 7-25:1,[41] to a redwood tree, which may have a 500:1 ratio. In case you were considering composting a family member, they have a 30:1 ratio, just perfect for microbial breakdown! Composters intrinsically understand this process, as they mix the right ratio of C:N materials to create a compost which will get hot, (nitrogen to fire up bacterial activity), but not too hot, (lovely cooling carbon).

Chemical pasture topping and herbicide applications provide an additional disruption to natural N function inhibiting N fixation. Herbicides not only disrupt natural N-cycling bacteria and create finer soil structures, but they also lead to the loss of 15-20% of applied nitrogen. Consider this when planning your spray timings and always include carbon with herbicides.

Often when we observe good density of legumes under trees or in pastures, there is an assumption that nitrogen fixation is happening. This may not be the case. Conditions must be optimal for the symbiotic process of nodulation and N-fixation to occur. In an invaluable symbiosis that occurs between plants and specific bacteria such as actinobacteria, Frankia, or rhizobia spp., the bacteria gain access to the plant roots, where they form small galls or bumps called nodules.

The symbiotic bacteria take sugars and minerals provided by the plant as payment, to then convert atmospheric N_2 into ammonia, which the plant uses to grow. This N is also transferred from plant to plant via mycorrhizae. If the legume nodules are not functional, this also points to possible issues with the free-living N fixers. Plants with healthy populations of both the N-fixers and MF, experience substantial lifts in health, gene expression and yield. [42] [43]

During the life-and-death processes which drive healthy biological systems, nitrogen goes through a variety of forms, before it can be withdrawn by plant roots. Bacteria have a remarkable ability to take (fix) and hold it in their bodies. If the soil foodweb has been compromised, through compaction or high soluble N applications, there is often lower predation from protozoa and nematodes.[44] Through this process, N becomes immobilised (tied-up), unavailable to plants. Picture bacteria as bags of fertiliser, sitting unopened in a shed. It takes an action from protist and nematode grazing to effectively open the bag up and release it to the plant.

Fungal biomass and activity, which is so essential for soil integrity and plant health, is also vital in reducing N losses.[45] [46] It's been calculated that the activity of fungi, such as mycorrhizae alone, can reduce N leaching by 40%.[47] The entire soil foodweb, from bacteria to nematodes, ants to caterpillars, play a role in the N cycle. Nitrogen is a valuable commodity traded between microbes and roots, and concentrated in the bodies of herbivorous insects, to be released when they poop or die. Indeed, all these organisms represent the power station, which barters and banks N, supporting optimal plant health and mopping up any "waste."

Degeneration of soil health with continued additions of soluble N, creates more bacterial soils, soil erosion and leaching, with waterways receiving the full brunt of negative effects. Research and producers around the world, are showing that high yields can be maintained, and inputs reduced through good management of soil, water, energy and biological resources.[48 49]

Legume Nitrogen Fixation

Dig a hole; do your legumes have copious amounts of nodules and are the nodules large?

When you pinch or cut open a nodule, is it pink or blood red in colour?

If the nodules are white- then they have not been fixing N. Time to put on your detective hat...

- Do you have low functional molybdenum (Mo) or cobalt (Co)? Both trace elements are involved in essential enzymes to metabolise N.

~ Is the soil already high in nitrogen? Have you been adding nitrogen or high rates of manures?

~ Is the soil saline, alkaline or highly acidic?

~ Did you inoculate your crop and is the inocula present in the soil? If nodules are green, the plant has been fixing and now the process has stopped. Ask why?

~ Have there been environmental stressors?? Cold soils will stop fixation in some species, as will drought and low sunlight.

~ Did someone just apply N? This action makes the natural N-fixers redundant.

7

Drinking It In

"Although the surface of our planet is two-thirds water, we call it the Earth. We say we are earthlings, not waterlings. Our blood is closer to seawater than our bones to soil, but that's no matter. The sea is the cradle we all rocked out of, but it's to dust that we go. From the time that water invented us, we began to seek out dirt. The further we separate ourselves from the dirt, the further we separate ourselves from ourselves."
– Tom Robbins, *Another Roadside Attraction*

Those who work with land, know well, the value of rainfall for the wellbeing of plants, animals and themselves. Most producers collect rainfall records, and during dry spells, or deluges, water is front and centre of every landowner's thoughts. All living organisms contain water in and around their cells; 70% of your body swims in water and for bacteria, 80-90%. Although some plants, such as 'the resurrection plants' and organisms like the indomitable water bear (tardigrade), who can desiccate their bodies to 3% moisture; all organisms require water to reproduce and fully function. Water is the giver of life and death, the driver for biological function, nutrient uptake, cell health and decomposition.

We are now exiting one of the most stable and accommodating climates in the past 11,700 years; these ideal conditions enabled the explosion of agriculture and human civilizations.[50] Climate is becoming increasingly variable and extreme across the planet, and food producers are the vanguard bearing the brunt of its effects. Modern agricultural research, with its desired values and recommendations, assumes production in these stable conditions. Instead, the regenerative producers are focused on farming into an uncertain future. With water quickly becoming one of the key resource issues of the 21st century, effective management of our underground workforce cannot continue being overlooked. Worldwide, the breakdowns in water cycles are becoming increasingly evident on the ground.

The increased political and activist interest in carbons role in climate change, ignores the elephant filling the sky. It's water vapour which is the most powerful greenhouse gas. It is impossible to talk about carbon without considering water; they are intimately coupled. Our moderate climate is largely due to water's amazing ability to absorb solar radiation and infrared heat. Without the buffering role of water in the atmosphere, global temperatures could be 33° C degrees higher. Water vapour's role in human-induced climate change, has largely been ignored by climate scientists, as human contributions to this pool were assumed to be minor.

The affable Walter Jehne, a former climate scientist with the Australian government, is ringing a warning bell that land management is having a far larger impact, than previously considered. Climate variability is being intensified by the 'humid hazes,' that are replacing rain clouds across the planet. Across the planet, fine dust aerosols are being kicked up from human activities such as; city sprawl, land degradation, desertification, fires and fossil fuel emissions.[51] These humid hazes arise from micro-droplets which form around the aerosols, which are too fine to settle and too electrically charged to develop into full-blown clouds. Jehne warns that this breakdown is a key driver to climate change. He estimates that every year, over 3.5 billion tonnes of dust aerosols, are ejected into the atmosphere. Our land management lies at the heart of how to reduce fine aerosols and put water back where it belongs, in the soil.[52]

Western Australians have undertaken a huge geo-engineering project, removing coastal vegetation, causing biodiversity losses and engaging in the irrational practice of chemically fallowing land. Trees and plant cover are essential in precipitation; most rainfall which falls on land is recycled back by plants through transpiration. Studies in other low rainfall cropping areas, have shown that the practice of leaving fields naked to the elements, changes rainfall dynamics, pushing the atmospheric boundary potentially 200m higher.[53] As a result of bare fallow practices, a large barometric high has been created over Western Australia. This, in part, pushes rains to the south and away from the wheatbelt. Following the deforestation upon colonisation, Summer rains dropped by 12%. Across the wheatbelt, Winter rainfall, when these Australian crops need it the most, has declined by 19% in the past 50 years.[54] Perth bears the brunt of one of the shifts with 50% less runoff now available to refill city reservoirs. Climate change is pointed to as one driver, with northern areas getting wetter.

Microbes play an important role in rainfall. Between 69-100% of precipitation is caused by the ice-nucleating bacteria that seed clouds, from families which include Pseudomonadaceae, Enterobacteriaceae and Xanthomonadaceae as well as *Lysinibacillus.*[55] Samples collected from clouds found over 30,000 different species of bacteria and fungi; a quarter of the bacteria and most of the fungi collected were actively metabolising.[56] Generally these species are found in the environment, living on, or in plants as symbionts, endophytes or pathogens. A type of rust, *Puccinia lagenophorae*, which ravaged wheat crops in the late 1920s, may possibly have had a role in altering climate dynamics, becoming one contributing factor to the 1930s dust bowl. Our farming practices have far reaching impacts impacting on living green cover and the vital microbes that create rain across the planet. Climate change doesn't help, but it's not the only culprit. Cooling the planet comes back to our land management choices; maximising green living cover,[57] increasing soil water holding capacities and fostering microbiology.

With extreme climatic events now a regular occurrence, 2017 was making the news for all the wrong reasons. 2017 marked record heat waves, cold, droughts and massive rainfall, and storm events causing unprecedented flooding and mudslides. There were large-scale catastrophic events in the Philippines, Africa, Peru, Nepal, India, the US and New Zealand. There was a lot of media attention around US cyclone Harvey in August, while at the same time catastrophic mudslides in Sierra Leone killed over a thousand people. A month later, flooding affected the lives of over 6 million people in Bangladesh. These impacts on people and food production are not solely because of climate change and deforestation; soils are degraded, they are no longer acting like a sponge. We have unwittingly created water-shedding catchments across the planet.

At the end of 2016, we were sampling around Auckland, New Zealand, working on farms just starting on their soil health journey. Soils on all the farms we sampled, had water infiltration rates measuring between 0.5-0.7mm/minute--that's a quarter of the conservative minimum. These beautiful young volcanic soils had good organic matter, yet all were badly compacted due to poor soil management. As a result, their water cycles were a bust. After a long dry spell, torrential rains hit the following March. In a one-hour period, Auckland city sustained 65mm (2.5") of rain, leading to sinkholes and massive flooding, with over 300 homes under water. Over 100mm (4") rainfall fell in a 2-day period, which is more than the month's average. Flooding in the city, can be attributed to human thoughtlessness in capping watersheds with concrete, compounded by the overwhelmed and blocked drainage systems. However, the toll outside of the city was as graphic. Many surrounding rural areas were completely underwater. Driving around rural areas that day, many main roads were closed and fast moving, dark chocolate brown waters covered the fields--our valuable soil is abandoning our lands and moving into the sea. They evacuated the township of Edgecumbe, on the East coast of the North Island of New Zealand, when flood-banks failed and inundated the entire town. This was after receiving three times April's average rainfall, in just three days.

Surely, you say, these are natural events and with extreme climate increasing, there's nothing we can do? Not so. Our natural capacity to absorb water has dramatically decreased. Healthy soils can absorb over 250mm (10") of rain in an hour. That's nearly 10 times the infiltration rate I had been measuring around Auckland. Poor soil management is impacting on society on a massive scale, yet it doesn't feature in most people's minds, including the politicians. With huge and mounting costs to society, a focus on soil regeneration is overdue.

Infiltration rates are measured in mm/minute. (25mm =1 inch). It's simple to calculate. Just divide the depth of water, by the time it takes to soak in.

It is critical that water soaks in faster than 2 mm/minute or 120mm/hour; any slower and you'll be losing water to run-off during storms; and under lighter rain, any water that falls, will just evaporate.

Take the example of one alumnus from the Ranching for Profit School. Agee Smith farms with his family from Cottonwood Ranch, which is located 2.5 hours north of Elko, Nevada and south of Twin Falls, Idaho. Making my way to the ranch was an adventure in itself, as I was towing a trailer, while avoiding cattle on the roads, kamikaze deer and potholes so deep, I'm sure they loosened my teeth fillings. Cottonwood also offers a tire shop service, as a testament to their road. This part of the country, is a cold, high desert, with meadows based at 6000 ft.

The Eastern part of the Ranch rears steeply up to over 10,000 ft, to form the jagged Jarbidge Mountains, one of America's least explored wildernesses. This mountain is the last remaining vestige of a violent caldera, what was once the Yellowstone mega volcano. Over 15 million years of intervening continental plate movement, the Yellowstone hotspot has settled to its current location, 300 miles to the northeast. The eruptions here blew thick layers of ash, suffocating prehistoric herds of camel, horses and rhinoceros, under 2 meters (6 feet) of ash. The geologic history is present today, with every step, shiny black pieces of glass catch your eye; the ranch is littered with volcanic obsidian flakes and chips. A visiting geologist pointed out to Agee that every flake found on the ranch, was once handled by a First Nation person. They traveled here to source the high-quality obsidian for flint and as they came down off the mountain, they chipped away the waste, to leave pieces to be discovered thousands of years later by a curious New Zealander.

If anyone ever wanted to find out where Santa Claus goes during his downtime, I'm sure I've found him holed up in Nevada. Agee's rosy cheeks, bright smile and his mischievous twinkle, belie the many years of hardship, he and his wife Vicki, have endured on the ranch. As a teenager in the 1880s, Agee's great-grandfather Horace Agee, head West. Originally, Horace ran freight wagons, with a contract to carry ore, before he met his wife, Etta Steele, the daughter of the way station at Ely. The Steeles needed help on their ranch, so Horace moved into the sheep running business. As their ranching success grew, they began to move deeper into the O'Neil's basin. The basin's namesake, the O'Neil family, had been notorious thieves, stealing cattle to sell to miners. The elder O'Neil and one son were caught rustling and were thrown into jail, before his 3 other sons broke them out. In the resulting gunfight, the father was killed, and their brother was wounded. The O'Neil's then moved to the basin, away from their mining customers. As a result, they needed to get smarter around how they obtained their cattle, stealing unbranded calves instead, before building up a more reputable cattle business. As Horace began to graze further into the basin, the O'Neil's frowned upon his choice to run sheep in the area and so they hired a gunman in an attempt to shift them off their land. Horace faced him down and the gunman fled the scene. In 1921, their feud came to a more honorable resolution, when Horace brought O'Neil out and the family settled quietly into the basin.

The tumultuous history was not over for the family. In the 1980s, with debt, high interest rates and a disease that wiped out their calf crop, the ranch fell into bankruptcy. Fortunately, there were others who also saw the beauty and potential of this vast wilderness area. The Smith family entered into a partnership with a grand vision, building an impressive guest lodge and turning the ranch into a hunting mecca. The ranch now offers dude ranch services, fishing and hunting, horse riding and fine evening conversations.

Agee's daughter, McKenzie and her husband, Jason, make a capable pair; the kind that came from breeding stock who had no trouble shooing off pesky gunmen. Jason also provides guiding and outfitting services for hunters and in Fall, McKenzie offers a service to pack out hunters' game, with her partner Jessica Mesina. McKenzie is a striking blond beauty. She reminds me a little of the mother antelope on their range, with their delicate tawny features that hide an underlying strength and resolve. When hunters come to the lodge and ask for help packing out, they often look at her twice as if to say, "Really? Is your husband going to help?" as if she wouldn't be up to the task. It's on my bucket list, to go up with her one Fall, to lend a less capable hand. To witness her diminutive power, as she carves up an elk to bring out of those intimidating mountains, or maybe just to see the look on those hunters' faces. Priceless.

In 1995, the ranch created a community partnership to address concerns around livestock impact on public lands, particularly around riparian health. Cottonwood Ranch manages nearly 40,000 acres. 1,200 acres are privately owned, and the rest is leased from the BLM (Bureau of Land Management) and USFS (United States Forest Services). At the turn of the 20th century, overgrazing occurred under excessive cattle and sheep numbers. While in more recent history, with the advent of barbed wire, the overgrazing continued, now by animals remaining on the same piece of land for too long. As a result, two thirds of their waterways, stretching 20 miles across the ranch, were in very poor condition, with sedimentation, unstable riverbanks and losses in diversity. Agee explains, "It had always looked this way, we turned cattle out in Spring, spent Summer putting up hay, then gathered cattle in Fall. This resulted in season-long grazing, which caused overgrazing on some places, like the riparian areas and over-rest on other parts of the land." By the 1990s, management issues had reached an impasse, that threatened the economic viability of the ranch and a public relations disaster for the agencies involved; Agee said, "It felt like our backs were against the wall."

The experience at Cottonwood, is an example of a 'wicked' problem; with multiple impact factors including climate, ranch management, BLM guidelines and multiple shareholders, family, hunters and fishermen and demands from the community to keep the environment pristine. Inspired by a class on Holistic Management, the Cottonwood Ranch Holistic Management Team (HMT) was born out of an agreement to explore the benefits on offer. The group involved 40 people, including 7 government agencies, private landowners, neighbours and the concerned citizens of Nevada. Holistic Management (HM) was brought to the US in 1979, by the now world-renowned, (or infamous, depending on your circle), Zimbabwe land stewards and educators Allan Savory and Stan Parsons. Parsons was raised on the largest dairy and pig operation in Zimbabwe, with a father who ingrained in him that agriculture needed to be profitable. Savory was originally a game officer who advocated a brutal decision to cull 40,000 elephants, on the premise that their high numbers were causing overgrazing. The culling did not reverse the land degradation and later Savory called the decision "the saddest and greatest blunder of my life." His realisation, that it is management, not animal numbers, that creates overgrazing and desertification, was the turning point in his life.

The HM method urges producers to integrate ecological, social and economic concerns, managing lands in whole systems methods which balance energy, minerals and water cycles. However, the approach is probably best known for its grazing method, which advocates a herd effect, like the impact of the wildebeest and zebra in Africa. This method of grazing, also referred to as MiG (Management intensive Grazing) or AMP (Adaptive Multi-Paddocks), focuses on high livestock densities, short duration grazing, with long periods of forage rest to maximize grass growth. The system requires observant and adaptive land custodians, as it is not prescriptive; instead, animals are shifted in response to the interactions between climate, land and life. Savory came under fire, for not scientifically backing up his claims. Dr Richard Teague, Texas A&M and others, are now collecting reputable, on-ranch data, to show the potential the approach can have for land restoration. As with any actions involving people, the results are as varied, as the personalities who manage the land and there have been great success and great failures from those following the guidelines. Fortunately, the Cottonwood crew are a tenacious bunch, with their livelihood on the line and a community looking over their shoulder, it drove them to show that the methods could reap rewards for land, agencies and their future.

The ranch supports a stunning array of diverse plant and animal species, from the privately-owned lowland meadows, to sagebrush foothills, mountain shrub, aspen, willow, alder, before progressing to mahogany and conifers on higher elevations. This diversity provides food and refuge for mule deer, antelope, elk, trout, spotted frogs, beaver plus a multitude of songbirds and the endangered sage grouse.

This diversity was one reason the ranch drew such a range of support from different agencies. The HMT sets biological plans each year, specifying grazing use and patterns for the upcoming year. They divided larger pastures into more manageable subunits, to ensure proper timing for plant recovery and to test and monitor the effectiveness of different grazing strategies. At the start of the project, cattle numbers were doubled and grazed using high intensity/short durations. They graze fields at differing times of the year to reduce over-grazing of different plant species. This timing is critical, as some species may be flowering at different times during a season. Focusing on management and grazing timing, was the biggest driver for change. Historically, under the BLM high country lease requirements, the land was to be grazed with lower stocking rates. By deferring their grazing, allowing longer plant recovery and then grazing later in the season with higher stock density, grasses were stimulated and allowed to seed, and they avoided overgrazing. They also implemented low-stress animal handling techniques and a technique known as range riding.

Range riding is a herding technique which encourages livestock to bunch together and move as one mob. In their terrain and with realities of raising a young family, McKenzie and Jason found a modified version of range riding worked better for them. They ride around their cattle most days, ensuring cattle are out of waterways and not creating stock camps around favoured areas. The concentrated movement of cattle, tramples grass evenly, disperses manure and urine and reduces the risk of overgrazing. The change in management, has resulted in a dramatic increase in plant diversity and waterway health.

Work in Montana, by the Western Sustainability Exchange (WSE),[58] has demonstrated that intensive range riding, not only improves grassland health, but it also significantly reduces the risk of predation from critters like wolves and to a lesser extent, bears, neither of which have shown up at Cottonwood. There is a view held in ranching communities, that range riding is uneconomic and unrealistic for family business. It becomes a chicken-and-egg situation: how to find the funds to pay a rider, before stock numbers can be increased.

Dr. Richard Teague's research, shows that in large paddocks, cattle typically cover less than half of large pastures. Using a range rider, can increase pasture efficiencies and increase stock numbers. Technology, such as 'virtual fencing,' is now providing management options for larger farms and ranches facing logistical challenges around fencing. Virtual fencing involves training cattle to electric collars, allowing land stewards to concentrate livestock and remotely move them without the need for fences. Compared to the cost of putting in more fences, reduced stock losses and the value in improving grass cover and soil and doubling or tripling cattle numbers, these numbers stack up.

Before trappers first arrived in the area, the waterways around Cottonwood Ranch were dense with beaver and their dams. Life has not been kind to beavers over the past 150 years; due to trapping, beaver numbers fell dramatically to just a few individuals. With the combined pressure from stock damage to riverbanks and Spring flash floods, their carefully constructed dams would blow out around their vulnerable edges. Implementing holistic grazing management, has given riparian areas a break, allowing the recovery of sedges, grasses and woody tree species like the alders and willows; beavers' favourite building materials. Changes in grazing produced smoother stream edges, versus the historic unstable and eroding cliffs. Where the ranges meet the lower meadows on Cottonwood, spectacular land changes are now clear. With these gentle stream edges and more woody vegetation, the beavers have flourished, and their dams can now withstand nearly all flooding events. The 2017 year was a catastrophic year, due to record snow-packs and then rapid warming. The warming created a massive flash flood which took out train tracks further downstream. Beaver dams blew out in their center, an effect never seen before on the ranch.

Under less intense conditions, water moves gently through the ranch, meandering and pooling up behind the dams. The water table has risen, and plant species dynamics are shifting. The tall sagebrush is now dying back due to wet feet and native grasses are returning, species such as wild basin rye, a grass so tall, that early explorers were astonished to report that "they could tie the grass over the tops of their saddles."

Urban planners are programmed to be prepared for possible spillover or flooding events; they aim to ensure water is removed as quickly from land as possible. These rushing, barricaded, channels with straight lines and high berms, are a false and shallow imitation of how water naturally moves. A regenerator's goal is the opposite; how can we support slow snaking waterways which dance during peak flow? For those living in environments free of beaver, building leaky weirs or adopting "natural sequence farming" as promoted by Australian farmer, Peter Andrews, helps to slow water down. I've visited properties who've found innovative ways to imitate the beaver, dropping in logs, old fencing materials, hay bales, rocks and yes even a concrete kitchen sink, into wash-out gullies and parched stream beds. The results can be remarkable. Instead of water rushing through and out, it slowly backs-up, soaks in and re-charges aquifers, as nature intended.

Visiting the regenerator, Martin Royds at Jillamatong, in NSW, was a breath of fresh air and joy for me after months of sadness and horror, at the state of livestock and land across Eastern Australia. Every inch of water that falls on Martin's place is captured and stored, before being filtered and slowly released over years. The water rushing across the surface of his neighbours land, disappears within a few meters upon hitting his fenceline. Martin sees his farm as an inter-connected ecosystem and the benefits are tangible. His once ephemeral erosion gully has been transformed into a perpetually flowing stream.

Through learning to read his landscape and livestock, he is trailblazing methods that create profitable production, in these crippled Australian environments. His approach uses a combination of holistic grazing, natural sequence farming and adaptive planning, which includes de-stocking before crisis points. He has constructed leaky weirs and contour ditches to slow water down, drawing it towards dry points. In 2019, he offered his valuable water to his local town, as rivers and dams in the drought struck area, have dribbled to a halt.

During my visit, we counted 12 different fungal fruiting bodies, including a woodland dotted with the black heads of *cordyceps gunnii. The* cordyceps family includes over 400 different parasitic fungi (entomopathagenic) which infect insect hosts, growing upon their rich protein bodies, before sprouting massive phallic structures from the hosts dead body.

The humble cordycep, were behind the controversial banning of 27 Chinese athletes from the 2000 Sydney Olympics. Used extensively in Chinese herbal medicine, some varieties are used for stamina and endurance, cancers, immunity and yes, libido. Locals in the Braidwood area have been willing guinea pigs in testing the efficacy of cordycep tinctures. Not only are water cycles being restored around Braidwood, marital relationships are too.

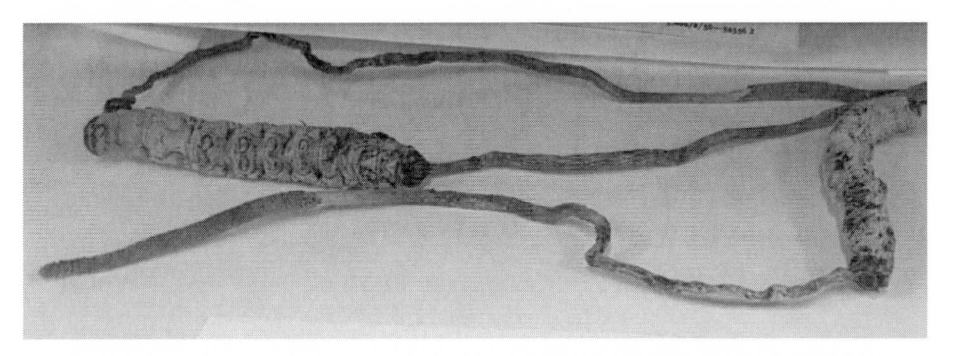

Image: cordyceps mushroom growing out of native New Zealand caterpillars.

As the water moves down towards Cottonwood Ranch, it passes through a series of irrigation ditches and dykes. The irrigation scheme was originally constructed in the 1890s, with hard sweat (and tears, I'm sure). Because of its age and the challenge of maintaining over 15 miles of ditches, some areas now receive an excess of water and others receive none. Since 1890, the lowland irrigated meadows were used intensively, cut yearly for hay and at times grazing over 270 horses. Agee's great-grandfather, Horace Agee, was suitably proud of his herd. They were originally bred as part of the Remount programme, providing horses for the US Army, until the late 1930s. These fine horses were a blend of Percheron, Arab and Thoroughbred, made for strength, speed and stamina. Over time, Quarter Horse bloodlines joined the mix, to create a fit-for-purpose mountain horse (one of whom I'm delighted to call my own.)

Add horses to the pressure of long-term hay harvesting, zero inputs, large pasture sizes and an irrigation scheme in decline, we have a perfect example from the recipe book on "How to Destroy Soil Health." As a result, the south side of the meadows, which form the hub of Cottonwood Ranch, hit rock-bottom. What does rock-bottom look like? These fields have become hydrophobic (quite literally afraid of water) and much of the ground cover, is dominated by rush known as 'Poverty Grass' ((*Juncus tenuis*). These south side fields are so bad, we renamed them "The Bronx." To grow any grass, the family has to irrigate constantly and areas where the irrigation misses, dry up and become crunchy underfoot. These areas attracted a massive invasion of grasshoppers; the land here is alive with undesirable livestock. With the insights from Holistic Management and Ranching for Profit, the family knew cutting for hay was unsustainable, so this practice was halted in 2012. By that time, there was barely enough grass to cut or graze.

Despite their isolated location, the family have a global outlook on life. The ranch regularly hosts some of the world's leading experts, running classes on grazing, holistic management, stock handling and biodiversity. People would often walk across The Bronx, discussing management strategies to correct the issues. Interestingly, in all their time battling with the issues in The Bronx, not one person thought to dig a hole, not one. Now I'm not saying that to imply other people are wrong, I'm saying it because the solution to so many problems, lies below our feet and under our noses, yet few take out their shovel to uncover any underlying issues. In my estimate, less than 5% of landowners would even dig a hole before they make an offer on a piece of land. Buying land is one of the most expensive purchases we make, and we don't dig to see what we're getting ourselves into?

Across the stream at Cottonwood, are the "North Side" pastures. Since clearing the sagebrush 30 years ago, these fields have been managed using Holistic Management grazing principles; short duration, high animal impact, followed by adequate plant recovery. Here, the irrigation scheme is newer and is working optimally; this means they can control their water, flooding one field at a time and allowing it to drain. With the smaller, easier to manage pasture sizes and water, grazing management has been close to optimal. These are proven ingredients in our "Soil Health Cookbook".

In 2017, we ran a class together with interested neighbours, BLM and USDA staff. We dug two soil trenches 300 feet apart, one in The Bronx and one on the North Side of the stream. What we found, was literally and metaphorically, groundbreaking. When Agee dug the trench in The Bronx, the top 1 inch of sod, peeled off like a carpet. Under this layer, a pale-yellow soil (or rather a lifeless dust), was bone dry. Surprisingly, two feet under the surface, we found a river of water, of which only the creeping root systems of the reeds could reach. Reeds and rushes are a clever bunch; they can survive in compacted or waterlogged conditions by transferring oxygen to their roots. Anytime we're in waterlogged conditions, beneficial fungi will die, so you'll commonly find plant species which are non-mycorrhizal. Over 80% of the plant families we found thriving in The Bronx, were sedges and reeds, lupines and some Redroot Pigweed (*Amaranthus*); these weeds are telling us the water cycle is broken and that beneficial fungal activity is impaired.

Water is a key element for production and the second triage point to pay attention to. In comparing the two soils, we assessed how effectively water was moving into the soil, using an infiltration test (see Appendix). I love these tests; they're cheap, super simple and incredibly revealing--and if your soil is in good shape, they're quick too! It's a great tool to compare different areas on your land and effectively track the effects of different management. Or, what I like to refer to as the "smug test," comparing your neighbors' results with yours, as you improve your management.

Visualize soil, as an apartment building, containing hallways, stairwells, corridors and living rooms. It's these spaces which enable soil to drink in any rainfall. A poorly structured soil will collapse, blocking soil pores and cause ponding and water to run off. And worst-case scenario, if a soil is hydrophobic, water will just sit on the surface and not absorb at all.

When we completed the test on The Bronx, it took over 24 minutes and only one inch of water went in. When we pulled the rings up, the water was only sitting on the top inch. Agee tells me that the ranch receives 10 inches per year. "Nope," I replied. "Sadly Agee, in these fields you're not even getting half of that." These soils are not functioning like an apartment building; they are a tarmac.

The trench in the North Side, was an absolute contrast to the one in The Bronx. It's one of the nicest soils I've ever seen. I tell you; it took a lot of self-control to not roll in it! It was a lovely dark brown, with a beautiful chocolate cake structure. Most of the grass species were Garrison Creeping Foxtail (*Alopecurus arundinaceus Poir*), an introduced cool season perennial grass, often found in moist areas. This palatable grass was growing over a meter (40") high. These deep, dense root systems had huge dreadlocks dangling down into the soil. The rhizosheath is where the action is; here there are billions of microbes, bacteria, fungi, nematodes and protists, all working away to support plant health. These are the guys that build the apartment building. These microbes are also the ones responsible for providing nutrients, water and essential metabolites and vitamins to the plant, our microbial bridge builders. We sent soil biology tests away to compare the two areas and sure enough the MF came back as zero in the Bronx grasses and optimal in the North Side. The Bronx also had poor biomass and diversity of all the other soil organisms, critical for plant health and production.

There are specific bacteria and organic materials which create non-wetting or hydrophobic soils. When these bacteria dry up, they protect themselves with a waxy coating. This is a massive issue in semi-arid, sandy soils in Southern and Western Australia. Over 5 million hectares have these non-wetting soils, costing the wheatbelt a fortune in lost production. At Cottonwood, an important aspect of our programme was the application of 2 kg/Ha (2 lbs/ac) of vermicast, as a liquid extract. There is a multitude of reasons to add vermicast: plant growth hormones, humic substances, nutrient solubilising bacteria and fungi, frost eating bacteria and more. More specifically in this case, the diverse microbes in worm castings include organisms which break down these waxy coatings, including bacteria from the genera Pseudomonas, Micrococcus, Nocardia, Corynebacteria, Bacillus, Arthrobacter and Proteus. Who better to come to the rescue than the elixir of life that comes out of an earthworm's butt?

The breakdown of the water cycle on Cottonwood, is not just about water repellency, but also water storage. We need a soil to operate like a sponge. As we covered in previous chapters, a large component of soil's ability to hold water, comes from the soil humus content. There was a visual difference in humus levels between the two soils and also as measured on a soil test. The humus levels in the North Side, were 7 times the level of humus in The Bronx. As a quick calculation, humus can hold 7 times its own weight, this means the North Side can effectively absorb and hold onto 50 times more water! The Bronx is in serious trouble.

Not enough land stewards are taking soil tests to assess what their mineral bank account is. Taking a benchmark soil test is valuable to assess whether you are improving or degrading your resources. As land stewards, we need to record this information, for ourselves and future generations to say, 'Yes, I did a great job as a steward.' If the tests show a decline over time, they can provide an early warning to act, before land turns into a situation like The Bronx.

At Cottonwood, we took soil tests which included North Side and The Bronx fields. At a demonstration to a group of scientists, they argued against the conclusions drawn. It was clear in their minds, that the striking differences in tests and profiles were not due to management, but different soils. In their opinion, The Bronx was "natural", due to soil type. This can be a common response, that degraded soils and landscapes "have always been this way." Many landscapes degraded quickly, after people colonised an area and without photographs and soil surveys, it's hard to argue things were ever different. In the brilliant book "Dark Emu," Bruce Pascoe argues that it only took 2-3 years for the brittle Australian ecosystem to collapse. To prove to ourselves, Agee and I saddled the backhoe to look at soils downstream on the North and South sides, just outside the hay ground fences. We found these soils were in good shape. We then sampled in the rangeland on the same soil type, above the irrigation ditches and separated only by a fence from The Bronx sampling area. This South Upper test is a pretty good estimate of what The Bronx would've looked like before cutting for hay and irrigating. Most people have been trained to see disturbed poor and unhealthy soils as 'normal', but that's just not the case. I doubt there is any ag ecosystem in perfect health, and most are far from any form of perfection.

Optimal soil structure results from a dance between microbes, minerals and soil type. Sandy soils will perform differently than a heavy clay soil, for

Soil carbon is a giant sponge

An increase of 1% carbon can increase soil water holding capacities by the following amounts:

- Soils less than 10 % clay – 20 to 30% increase
- Soils 10 to 15 % clay – 10 to 25 % increase
- Soils 15 to 20 % clay – 10 to 18% increase
- Soils > 20% clay – about 10% increase or less.

The USDA determined the water holding capacities of organic matter (OM). Depending on soil type, a 1% increase in OM in a loam soil, increased water holding by 187,000 litres/Ha or 20,000 gallons an acre, (calculated as water holding capacity down to 30cm depth in a loam soil). Now this result is based on one moment in time; with drying and wetting cycles varying through the year, it may be 3 times this amount-- nearly half a million litres/ha/year, or 43mm of additional stored water.

instance, partly due to particle size, partly due to electrical charge. This particle size influences how well a soil can hold onto moisture and nutrients.

Imagine you have a teaspoon of soil and the world's tiniest measuring tape. With this tape, picture measuring the surface area around every grain of sand and you'll come up with a figure. Now do the same with a teaspoon of clay, measuring with your teeny tiny tape in between each plate of clay. If you could flatten the surfaces of all the grains of sand, it might cover the surface of a small dining table. In comparison, when we flatten out all the plates of clay, the surface area could cover at least a football field, just from one teaspoon (depending on the type of clay.) That's mind boggling!

When you take a soil test, most results will report a number for CEC; this is shorthand for Cation Exchange Capacity. CEC is a measure of particle sizes—which you got with your teeny tiny measuring tape—and their negative charges or their capacity or ability to attract positively charged ions. These positive elements are called cations. The base cations are calcium (Ca+2), potassium (K+), magnesium (Mg+2) and sodium (Na+). The acid cations are hydrogen (H+), aluminium (Al+2), manganese (Mn+2) and some trace elements. Acid cations are so named, as they decrease soil pH and make it more acidic, while the base cations lift pH, making soil more alkaline. This is important to note, as many people only consider lifting pH or high pH in relation to Ca; however, it might be due to the other base cations, or even from soil microbe exudates (poop, puke, sweat and spit).

Conceptually, the CEC test is a measure of the size of your bank account. Imagine a low CEC (below 5) as a piggy bank, these are the sandy soils, low in organic matter, humus and clay. While a high CEC soil (over 25) is often high in clay and/or organic matter. More like a bank vault, these high CEC soils have a large ability to store nutrients and water (as well as pesticides) and typically are more fertile than low CEC soils. Low CEC soils are more prone to leach nutrients, insecticides and pesticides into waterways.

The surface of these particles and soil organic matter, carries a negative charge. This difference in charge is the reason why sand has difficulty in retaining water and nutrients and why clay soils can get so wet and sticky. Imagine walking over a building site full of clay-- by the time you cross your boots are literally up on platforms. This is because of the negative electrical charge; soil is all about energy! It's why you want a nice mix of different particle sizes in your soil, the stuff gardening folks get all excited about called "loam." Unless you happen to be in Australia, where they call their white sand: 'silver loam.'

Consider that the size of your bank account are traits your soil was born with. However, with a focus on building humus and organic matter, you can have a positive influence on your CEC reading and how your soil functions. In some soil, CEC can decrease with time, through acidification from fertiliser, losses in organic matter and natural activities such as water logging and leaching.

It's also important to consider that different labs use a variety of methods to measure CEC. This means you can get different results from different labs. With all testing, we recommend you choose a good lab and stay with that same lab, so you can track changes over time.

The soil mineral tests at Cottonwood confirmed that management and irrigation practices had effectively leached out many of the minerals in the soil. These graphs show measurements of the Cation Exchange Capacity (CEC) and Base Saturations. CEC levels are low in The Bronx (7.4) and are higher in the North Side (11.8) and South Side Upper (12.9). South Side has eroded/leached out much of the finer particles and humus over time - which reduces the CEC.

The final breakdown of organic matter, humus, has an even finer structure than clay. Think about when you see someone cultivating a field; all that fine dust you watch blowing away is the most valuable resource: humus.

It's all very well to know how big your bank account is or isn't. More importantly, you want to know how much money you have in your account. The Base Saturation (BS) reveals how full your bank account is with the base cations: Ca, Mg, K and Na. Our goal is to have 70-85% BS, with the remaining space for H and Al. There has been some controversy around using BS for any diagnosis or recommendations. However, we consistently find there are some aspects of BS which will put a drag on any property, particularly when Ca is low and/or Mg, K, Na are high. Base Saturations are not 'perfect' numbers to chase, they are informative around structure and potential and over time as biological integrity is restored, we see many of these numbers change.

These pie graphs (following page) show the CEC results for the Cottonwood fields. Notice that the South Side Upper has a bank account FULL of base cations. This is called 100% base saturation. This is common on many semi-arid rangeland and desert soils we've tested. If you have 100% BS, don't let the salesman sell you more cations (they will always try), as there's nowhere for the minerals to stick! Both South Side Upper and North Side have excellent base saturations of Ca and Mg, are low in Na and have an excess of K. The base saturations in the South Side are all critically low. When these elements are out of balance, there are serious implications for soil structure and plant/animal performance.

The South Side Upper and North Side both have an excess of K. This is due to the parent materials in the area. Excess K can create tight sticky soils, with implications for soil compaction. However, when well-managed, this K provides ample nutrients for plant growth.

If you compare the graphs for South Upper with The Bronx, you can see clearly, just how much of the minerals have been exported or significantly leached out, leaving 57% of the soil bank account empty and full of hydrogen instead. Nearly 50% of all the cations have left the building! This loss has brought Mg and K down into a more manageable range. This is a good thing, as it can help fast track soil structure, once the low calcium has been addressed.

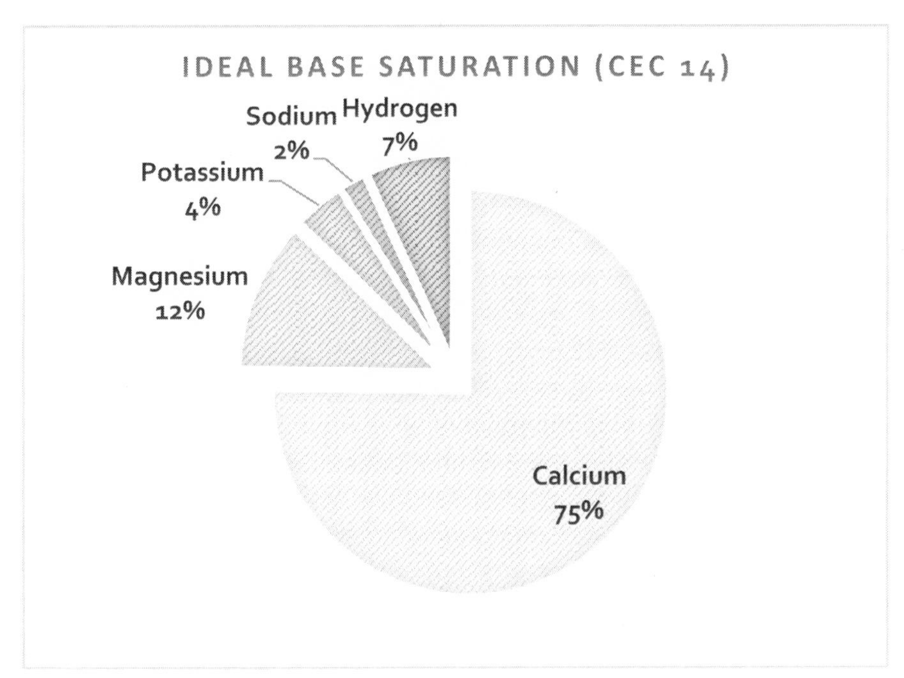
IDEAL BASE SATURATION (CEC 14)
Sodium
2%
Hydrogen
7%
Potassium
4%
Magnesium
12%
Calcium
75%

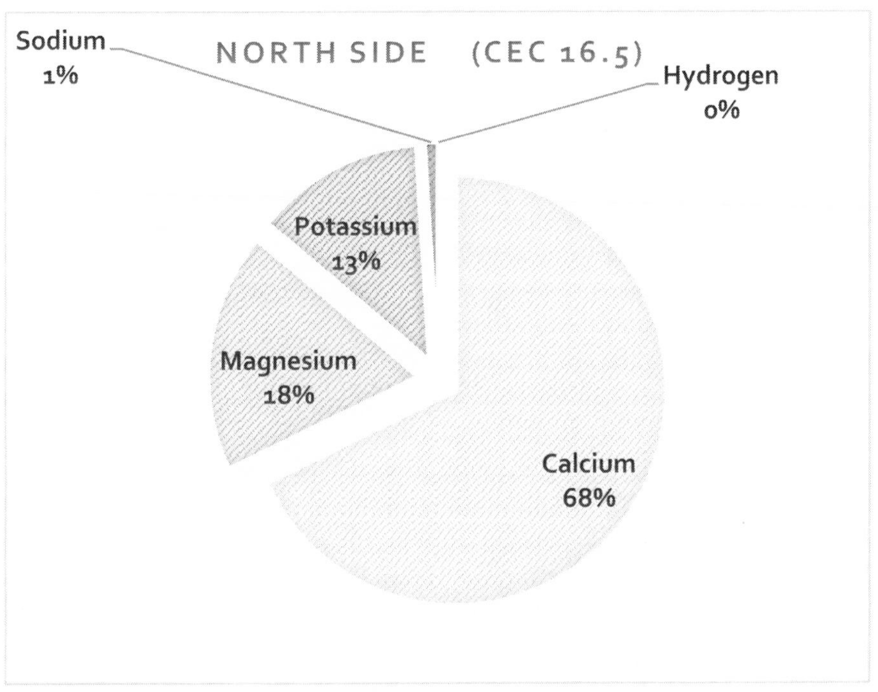
NORTH SIDE (CEC 16.5)
Sodium
1%
Hydrogen
0%
Potassium
13%
Magnesium
18%
Calcium
68%

SOUTH SIDE BRONX (CEC 7.4)

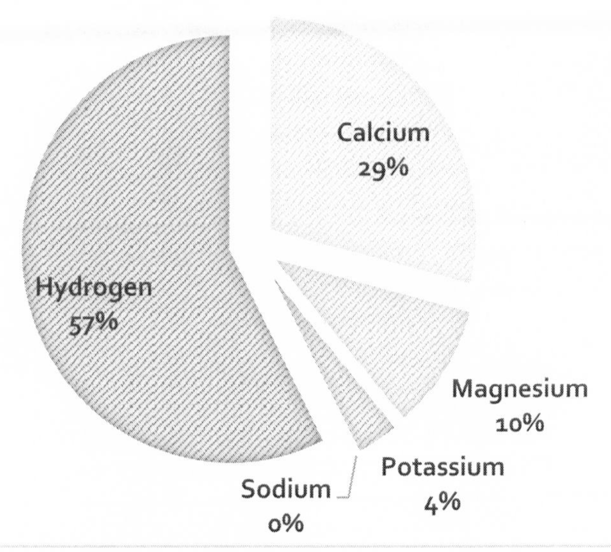
Calcium
29%
Hydrogen
57%
Magnesium
10%
Potassium
4%
Sodium
0%

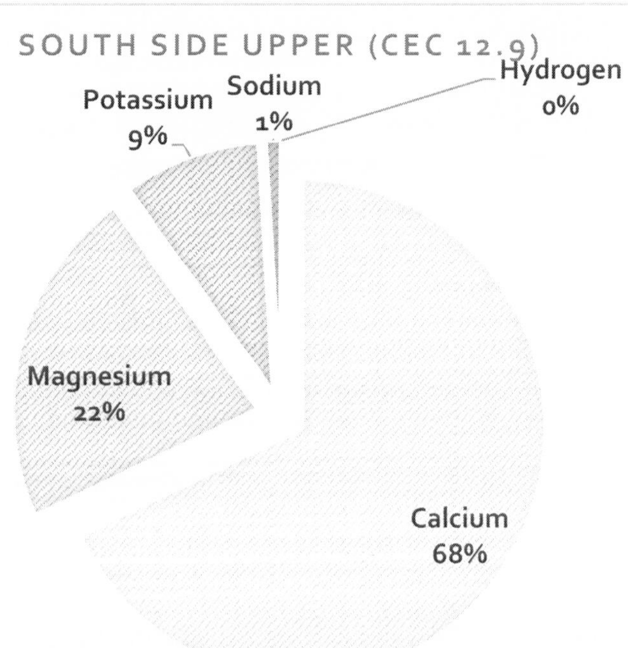
SOUTH SIDE UPPER (CEC 12.9)
Hydrogen
0%
Potassium
9%
Sodium
1%
Magnesium
22%
Calcium
68%

Calcium (Ca) is a foundational element in regenerative soil programmes. Why do we want Ca? Think of its role in your body; it's in every cell wall, tooth and bone. It clots your blood, supports muscles and nerves and supports your overall health. Ca is the foundation for human and animal health, similar to its role in the soil. The world-renowned biological agronomist, Gary Zimmer, calls calcium "the trucker of all nutrients," for its role in the plant. Any compromise in Ca is a compromise in soil structure, biological activity, plant performance and health. Key, in healthy and productive soil systems, it is involved in cell wall structure, cell membranes, drought resilience,[59] growth and pH balance.

Lifting Ca availability in soil, helps to lift plant, crop and animal health through the production of calcium pectates. Pectin's are the glue that hold plant cell walls together. These lift Brix, metabolizable energy, fruit quality and milk and meat production. The Ca pectate also keeps forage greener for longer, helping to overcome the drop in Vit A, which arises as plants brown off or become 'rank.' Through adequate Ca, we can lengthen the growing season for quality forage production. For fruit producers, this is quality and storability.

Microbes and worms love calcium and their populations flourish after application. For those of you with soils low in Ca, you will discover biological stimulants are ineffective. There is no point throwing compost teas and extracts onto a soil without good foundational Ca. Look to the examples around the world of success with extracts. These properties all have reasonable or high Ca levels. Ca is critical for the success of a regenerative soil programme.

With low BS Ca (below 50%), actions need to be taken to remediate. These low calcium soils are common in many high rainfall environments, with long histories of poor management. New Zealand and Californian soils can exhibit the impact from low Ca, with acidic, compacted soils as a result.

We have worked on properties where BS Ca was extremely low (<15%). On these soils, it was uneconomic or unfeasible to address the issue with lime applications. Through aerial amendments of trace elements and feeding microbes, the BS Ca on these properties lifted by at least 10%![7] In just one year! Measuring changes like this gives us confidence that our soils are headed in the right direction. This is how soils are meant to be built--with

[7] Soil test your subsoil and see if calcium is sitting down there.

worms, roots and microbes bringing minerals from deeper in the profile to the surface. If you're not sure if you can rely on this process, test your subsoil to see if there is Ca deeper into the profile. If not, Ca applications will be beneficial in your transition.

Base saturation Ca gives us an indicator, however, having adequate Ca does not always mean it is *functional.* Calcium's tendency in low biological systems, is to move down through the soil profile. Calcium's availability is limited by moisture, low boron, compaction and biological activity, particularly fungi. This is why adding fungal carbon foods: straw, woodchip, humates and fish oils; are so effective at lifting Ca availability. At Cottonwood, we knew from our visual observations, backed up by biological testing, that fungal activity was close to zero. Beneficial fungi are essential in retaining and releasing calcium oxalate to plants.[60] Poor fungal activity equals poor calcium mobility to the plant.

Take note, if you are applying calcium, always, always, always, feed fungi when you do. Many farms around the world are trying to correct pH and low calcium issues by adding truckloads of lime, with only temporary results. In extensive landscapes, I prefer to "tickle the system" with fine or liquid limes. Fine limes act as a catalyst and can kick-start biology, drawing Ca from deeper in the soil profile. When using these fine products, a little goes a long way. Usually, we use between 10-20 kg/Ha (10-20 lbs/ac). There is no need to use hundreds of kilos of fine products. You're just shocking the system. Think of these catalysts, like the lighter fuel in a BBQ, there's no need to overdo them--you don't need to lose any eyebrows in the process! These liquid limes can be made following the old recipe for whitewash. You'll need an agitator in your spray unit or the ability to emulsify.

One caution with calcium: as with all minerals, there may be short-term antagonisms which tie up or release other elements. As Ca flocculates and opens up soil, adequate Ca will improve water efficiencies, reduce compaction and lift production. During the flocculation process, minerals or even pesticides bound inside clay plates or compacted zones, become released. One visual clue that this is happening, is the appearance of more broadleaf weeds in the first few years in your soil programme. Don't panic. Often this is potassium becoming more available. It will settle down again. A group of NZ dairy farmers learned a costly lesson from applying both solid and multiple liquid lime applications in the year. As a result, potassium flushed to the plants. This imbalance resulted in metabolic problems and over 160 cows died. The counterbalance for excess potassium in feed?

Sodium. Providing salt licks to these animals could've saved their lives. Calcium is essential, but like everything in life, moderation is key.

There's another good reason for applying calcium at Cottonwood; research on Australian non-wetting soils found that applying limestone resulted in a 10-fold increase in wax-degrading bacteria.[61] A focus on lifting calcium across these irrigated grounds, is essential to meet the ranchers' goals: increasing resilience and palatable grass species, optimising water and turning profitability around. Most fortunately, a limestone quarry is located near to the ranch; otherwise, it would not stack up to apply solid lime. In that case, liquid lime would be more affordable, however that would need to be applied more than once.

With regards to the hayfields at Cottonwood, grazing alone, is not going to address a soil in this condition. There's just not enough pasture production to get the impact and trampling required and as the pastures are such low quality; they negatively affect animal health and performance. On low organic matter soils, techniques like bale grazing (the practice of bringing in large round bales to feed cattle and add organic matter to the soil), can be very helpful. However, this method can still take many years and a multitude of bales to address the underlying causes, particularly with the low calcium and water repellency creating a massive drag on the system.

The recommendation to rehabilitate The Bronx, was designed to jumpstart the system. A diverse cover crop was sown with dried vermicast and seaweed in their fertiliser box. The mechanical action of the drill, effectively opens the top inch of the hydrophobic layer, increasing water efficiencies immediately.

Cottonwood Ranch is a unique example of how management can either degenerate or regenerate land. The hayfields had become an area that they believed would never return to health. Degradation of landscapes has often taken decades, if not centuries to occur. Regenerating soils isn't instant, like cooking an egg, however, there are actions we can take to re-boot certain soils. The Bronx soil required a different course of action, firstly by restoring water cycles, by tackling water repellency, stimulating biological activities and addressing low foundational minerals--particularly calcium. These practical places to start, had already been identified by the Smith family: addressing the poor functioning irrigation systems and reducing pasture sizes. The future for these fields, is now looking much brighter, with whole scale regeneration actions and irrigation systems to be updated in 2020.

8

Break It Down

"Decline is also a form of voluptuousness, just like growth. Autumn is just as sensual as Springtime. There is as much greatness in dying as in procreation." — Iwan Goll

In a world devoid of its army of decomposers, we would rapidly be overcome by debris and despair. Microscopic bacteria, fungi, protists and insects of every size are the critical clean-up crew, digesting organic materials and keeping our world turning. They have an extraordinary ability to break down even the most resistant, radioactive and toxic of materials, transforming them into bio-friendly foods for other organisms. It is essential that decomposition cycles on your land are performing well. Without good decomposition, nutrient cycles and profitability will grind to a nasty, stomach churning halt. This is the fourth place to look to in our triage. Decomposition is the core of your soil's digestive function. Does your soil have indigestion, gas, constipation or diarrhea?

Across the world, cropping farms are now stalling under the weight of stubble that just won't break down, often under the chemical fixes prescribed to the no-till generation. It's becoming common, to find stubble remaining after three or even four years following harvest. In many cropping regions, producers have resorted to burning this oxidized material, as even a starving sheep won't stomach it. A rancher once showed me photographs of their monitoring transect; in every photo for four consecutive years, the same cow pat featured center stage. These are the landscapes of constipation.

I love the statement: "Bet you can't solve this one." Not only does it get my gung-ho competitive juices flowing, but it also means I might be about to learn something new, or potentially make the difference to a farm or ranch. A few years ago, while on a Northern Montana ranch, I had just this statement posed. Their irrigated fields were full of humps and hollows. When flooded, the low points dried up and turned brown, while the higher areas were lush and green. Very cryptic, areas which should be moist and green were dead. A quick dig revealed over 7 inches of thatch in the low points.

Thatch is the result of a build-up of organic material around the base of living plants. This layer forms when plant materials accumulate faster than microbes can break them down. There are some plant species which naturally form thatch, such as Smooth bromegrass (*Bromus ineris Leyss*) and Kentucky bluegrass (*Poa pratenis)*. Thatch is distinct from peat or muskeg soils, which naturally arise when organic materials accumulate in anaerobic, boggy, wet and acidic environments. Conditions for thatch formation, can include poor microbial activity, low pH or alkali soils, environmental factors and poor management. Practices such as heavy nitrogen applications, pesticides, compaction and/or over-watering can all contribute to thatch formation. Rather than growing roots deep into the soil, new roots start to grow into the thatch layer. When conditions turn hot and dry, the shallow-rooted plants struggle to survive.

Adaptive Multi-Paddock (AMP) and other holistic grazing practices place a strong emphasis on building up leaf litter. In long season, temperate climates, when using these grazing techniques, the litter won't build up; worms, insects and microbiology actively break the grass materials down before the next rotation. This concept of litter was developed in brittle environments that have bare ground. Here, it is essential to build litter or some kind of ground cover to protect soil surfaces from wind, sun and destructive raindrops. In less brittle environments, this service is provided by a green living plant.

There are six major ingredients required for microbial digestion: air, water, sugar, calcium and a little phosphorus and nitrogen for energy. With compaction, low Brix, poor calcium and low biological activity, the soil-gut digestive system shuts down. Through building our soil-gut health programme we, can restore digestion. For cropping and horticulture, we can kick-start digestion, using inputs and on rangeland and extensive stations, we can optimise decomposition through livestock management.

Good airflow will ensure anaerobic digestion is avoided. Without oxygen, the anaerobic bacteria move in, releasing volatile organic compounds (VOC) as they feed. Like a seedy drug den on the wrong side of town, these bacteria start to push their fermented and pickled by-products which include alcohols, formaldehydes, methane, carboxylic acids, esters, ketones, sulphides, terpenes and other organic acids. Your nose will tell you when many of these microbes are at work, as they releasing a 'farty', rotten egg (sulphur) and ammonia smells (nitrogen), as well as carbon dioxide and methane (more carbon).[62] These smells are a good indicator that you're losing your valuable nutrients, carbon and beneficial microbes. The foul stench of hydrogen sulfide produced in waterlogged conditions, leads to a variety of chemical reactions which produce metal sulfides. These include the insoluble manganese, which leave black flecks in soil and the ferrous sulphides which create the dark black of brown sludge in wetlands. As the waters recede, metals, such as iron sulfides, react and become iron oxides, literally rust, giving soil, distinct orange rusty mottles. If soils are anaerobic for longer periods of time, they give rise to the gleyed soils, with their blue and grey mottles. Some VOC are visible to the naked eye. In days gone by, shallow burial sites gave rise to stories of ghosts wafting through cemeteries due to the release of the bioluminescent gas, phosphine (PH_3). As a result, we learnt to bury our dead deeper.

Visual signs of these VOC, include shallow roots, which shear off abruptly along a horizontal line. These roots cannot live and breathe in these fermented layers. These areas will have lower quality growth, increased stress and reduced resilience during dry and hot conditions.

If thatch and undecomposed layers are over an inch deep, problems start to arise, as air flow is compromised. As a result, the conditions for disease set in, with insect pests, water repellency and nutrient breakdown. This creates a vicious downward cycle that can be hard to interrupt. With seven inches of thatch, the Montanan fields were in a world of hurt! This thatch has similar dynamics to The Bronx at the Cottonwood Ranch. Bacteria on the lignified root and grass materials, form water resistant waxes, repelling the irrigation water. In compacted soils or those with deep thatch layers, this is the time a mechanical intervention can be incredibly beneficial. Aerate or rip one inch deeper than the thatch layer to kick-start the system, by introducing air. Down these rip lines, you can use the same process as the horse stud at Lindsay Farms by dripping humic acid and sugar/molasses. In horticulture, managing post-harvest digestion, is critical to ensure diseases do not breed in between the growing seasons. Applying soil and leaf drenches with calcium-based sprays provides a hygiene clean-up. There are many commercial digestion products, or you could also make your own using a lactobacillus serum.

All who work on the land are aware of the vital importance of water to grow plants. Your underground workforce also has an essential need for moisture to provide fundamental life support. Making the most of every drop of moisture is key to addressing hydrophobic conditions, infiltration, capturing dew and keeping soil covered.

In frozen climates, where Spring melt provides early irrigation, these fields stay cold and wet for long periods of time, which slows biological processes down. Having well-aerated soils, with increased biological activities, can lift soil temperature and support growth earlier in the season. In hot, arid environments, soils can remain dormant for years, until life-giving rains arrive and biology springs to life. In these drier climates, where water is limiting, bio-stimulants can be applied during heavy dewfall, in the late evenings. If you are not using bio-stimulants on dryland areas, providing probiotics to livestock will ensure their manure breaks down faster. Manure will decompose quickly when using probiotics in troughs or on supplementary feed, or with free choice raw humates offered in tubs. Beneficial microbes can be supplied directly into troughs through living plants' mats, commercial probiotics, quality compost. Some producers leave compost in troughs permanently, potentially creating anaerobic conditions. If you are going to do this, leave a sack of good compost in a trough for 3 days maximum.

For those who have the benefit of irrigation, be that flood or overhead, you can shortcut application processes, by applying biological stimulants directly through fertigation. Producers are adding a variety of inputs such as; fish, molasses, sugar, seaweed, humates, vermicast or digestion products straight into ditches, with mixed results. If using ditches, efficiency can be increased by dripping products into the head gate. Apply with the first irrigation of the season, particularly with oily products like the fish, to get a more even spread.

A regenerative system encourages the workers in the soil to do just that ... work for you. One of the most important and visible members of your underground herd is the earthworm. These are the ecological engineers who tunnel and mix through the soil metropolis. In the process, they neutralise soil pH, leaving calcium, nitrate, mucus, microbes and quorum signal laden tunnels. These channels create pathways for plant roots, oxygen and water to pass. Earthworms are the planet's great alchemists, sucking up raw ingredients, combining minerals with enzymes and metabolites to produce rich, plant-available foods. In my mind, nobody builds soil better than these undervalued professionals. They have been tirelessly toiling away, building soil in the same manner as their ancestors 600 million years ago. In the magic that is the natural world, worm castings are perfectly designed to support optimal soil and plant health. Vermicast has wide benefits for soil health, increasing water holding capacities, soil texture and nutrient retention and availability. For plants, vermicast improves seed germination, plant health and production, beyond the wildest dreams of the synthetic NPK—all at much less cost to producers and the environment.

The castings they excrete, contain a diverse community of microbes, vital in supporting plant health and yields. These microbes deliver a wealth of soil contributions. To give just a few examples of their many benefits, the microbes in worm castings; prevent disease, reduce plant shock, eat frost-forming bacteria, stimulate mycorrhizae, break up waxy coatings and digest the hard to break down stuff. They work hand in hand (or rather poop to root) symbiotically with plants. In the field, applications of solid worm castings can provide measurable benefits for years.

Although it may seem that worms prefer living in fertile soils, it is in fact the worms themselves that make the soil fertile. They may be present in large numbers, attracted by the vibrations and heat created by feeding bacteria, one of their favourite foods. Hence, high earthworm numbers are not always an indicator for a healthy microbially balanced soil. As a general rule, low worm numbers in temperate sites are an indicator of poor soil management. Chemicals such as Lindane and 2, 4-D will kill earthworms. An average soil needs to contain at least 25 worms per shovel, which equates to 2.5 million/ha (one million/ac).[63] It is not uncommon to find counts less than 10 worms per square spade on chemically managed and cultivated farms, whereas well-managed landscapes may have over 70 earthworms. Numbers will depend on temperature, moisture levels and soil type. Worms are surprisingly tough and are being used to remediate contaminated mine sites.[64] They are powerful tools in the biotransformation of heavy metals, synthetic dyes, toxins and nanoparticles.

Despite their ability to remediate these toxins, when it comes to plastic, our other modern gift to future paleontologists, worms are out of ideas. Research just out, reveals that soils containing micro-plastics impacts on plant yield, soil structure and stunts the growth of our friend the earthworm.[65]

Digging a hole reveals more obvious clues when digestion has stalled, like thatching, 'bad' smells, gray oxidised layers on the surface or sharp changes in colour between topsoil layers.

The OM test includes all living materials in soil, not just humus and carbon. This measure may also contain thatch, plant roots and dead organisms smaller than 2mm in size. If your soils contain high levels of undecomposed organic matter, it may give you a false high reading. Humus and glomalin tests give a better picture around microbial conversion of the raw organic materials into more stable humus forms. However, these tests are less common and more expensive.

You can also request a carbon to nitrogen (C: N) test to help identify if your carbon is functional. In healthy soils, with a good decomposition cycle, this test will read a C: N ratio between 10:1- 12:1. If your ratio is lower than 10:1, you have an excess of nitrogen in the system burning your carbon up. A low C:N ratio can correspond to an excess of nitrogen in the system, rapid turnover of organic materials and burning up of carbon.

If the ratio is above 12:1, your soil decomposition cycle is stalling. This can lead to a build-up of thatch and a slower breakdown of manure and plant materials. As microbes work to breakdown C, they will rob N from the plants. High C: N ratios relate to the following combinations: mineral imbalances, low pH, low or imbalanced biological activities. These are the constipated soils.

This C: N test is a crude indicator that organic materials are converting to humus in the soil.

My first adventure with composting worms began in Dunedin, New Zealand's southern student city. A friend kindly exchanged a handful of tiger worms (*Eisenia foetida*) for a bundle of tomatoes and broccoli. The worms made their new home in a plastic box and were fed food waste from my student flat (or what American's would call an apartment, not a word that aptly describes this sunless and damp hovel). Unfortunately, the cold and exposed environment, resulted in a miserable death for thousands and I forgot about compost worms for a few years. Until my father's opportune stumble across an advert for the 'deceased wormfarm estate.' Through the good fortune of many brilliant worm farmers, including Thelma and John Williams at Pottsbury Farms (NZ), I created a fungal diverse vermicast. Over the following decade, I became the local and somewhat reluctantly named, "worm lady," delivering household waste minimization and a "worms in schools" programme.

I fed my worms a blend of pig and horse manure, avocado pulp, cardboard and white wood chips dusted with fine lime and the occasional light additions of rock phosphate and seaweed. Worms themselves are happy to feed on rich food sources and you can produce a vermicast product with only vegetable scraps. Cattle manure is a brilliant feed source for worms, but if used in isolation, the end-product becomes bacterially dominated. Through incorporating wood chips, the fungi move in and a wonderfully balanced product can be produced. Most commercially available vermicast is bacterially dominated. Visually this vermicast has a fine texture and if you leave it out in a seed tray in your garden, it will sprout primitive weed species. Ask a supplier for their biological tests. Bacterially dominated worm farms also release more liquid, or leachate, from the bottom of the bin.

The perfect worm farm makes no liquid. This information always comes as a shock to home gardeners who have been diligently and proudly collecting this leachate for years. This liquid is a combination of the undecomposed food wastes dribbling through the worm castings. It can contain diseases and nitrates. The production of liquid is telling you that your worm bed is bacterially dominated, and more carbon-rich materials are needed, such as cardboard and wood chips. Good quality commercially available worm tea extracts (vermiliquid) are extracted by flushing water through the finished vermicast. Full of the concentrated quorum signals and dormant micro-biology, commercial liquids are a lovely chocolatey or golden colour. The chocolate colour in vermiliquid is the humic substances and the golden colour is due to fulvic acids—both essential bio-stimulants and plant health promotants. These products are stable for up to a year and release no smell.

Just as it is for our soil programmes, we want to foster as much diversity into our composts and vermicast as possible. A diversity of inputs, including duff from native forests, rock dusts, rock phosphate and seaweed helps to feed diverse beneficial microbes, such as phosphate solubilising bacteria, purple non-sulphur photosynthetic bacteria or calcium loving fungi. You can alter the feed sources for your worms to produce an inoculant specific for your soils.

I began to create different beds of vermicast, depending on the different end users: a high-fungal, high woodchip product for orchards and a balanced product for pasture. An early experiment opened my mind to the potential for vermicast. I planted a single passionfruit plant into one 10 kg (22 lbs) bag of vermicast, under my stairs in a concrete alcove. Over the next three years, a triffid vine clambered over the railings and produced copious fruit. All from one small bag of vermicast! Compared to synthetic fertilisers, vermicast benefits are far broader and longer lasting without the unintended negative consequences.

The early results we saw with avocados, wetland species and pastures, rivalled any packaged fertiliser product and I'm not alone in that sentiment. Professor Clive Edwards at Ohio University stated in 1995, that "Vermicompost outperforms any commercial fertiliser I know of." The work by Dr David Johnson is showing this effect on yields with his BEAM bioreactor methods. This method incorporates worms to complete the composting cycle and, as he has shown, a little goes a long way.

Many of my soil programmes include vermicast as a kick-starter for all the previously mentioned benefits. As extracts, we are using it at miniscule rates of ½ to 2 kg/Ha (lbs/ac) and seeing big returns aboveground in plant density, diversity and quality. What I have found fascinating, is how semi-arid landscapes respond to low rates. In biologically active environments, that already have good organic matter and rainfall, higher rates of worm extracts may be required. In New Zealand, we had been using 700 kg (625 lbs/ac) of solid vermicast in horticulture and using 20 kg as a slurry on sheep and beef farms.

Delivering workshops around the benefits of soil health took me further afield. As a result, I stopped spending any time in my own garden. One day, as I pulled into my driveway, I spotted the recognisable silhouette of the tall, suspender-belted, cowboy-hatted, Polyface Farmer, Joel Salatin. He was standing in the pathetic remnants of a failed vegetable garden. It was a moment of horror and an amazing motivator for action! I had 20 cubic meters (22 cubic yards) of vermicast stockpiled and intended for the farm. The next day, I hired a rotary hoe to prepare the ground and spread the vermicast directly on top, 300mm (12") deep.

With motivation still raw in my mind, frantic planting ensued; fortunately, finished vermicast will not burn even the most delicate plants. Vermicast contains hormones, enzymes, vitamins and antibiotics to reduce plant stress and shock, as well as organisms that eat many common root and plant diseases[66]. Over the following years, this garden produced delicious, high Brix, nutrient-dense foods. These days, I've found myself living permanently on the road with my horse and trailer, which I love, however, gardening has become a dream relegated to a time when I will once again have such things as walls, doors and a roof.

One fascinating aspect with applying vermicast, is that it attracts earthworms to feed and breed. Composting worms and earthworms do not live in the same environments. There is a huge diversity of different earthworm species which work to different soil depths, loosely grouped by how they function in soil: epigeic, anecic and endogeic. You will find compost worms in dung and deep litter, that aren't official "earthworms." They can, however, travel through the top layer of soil, looking for rich food resources; these are the epigeic (greek for "on the earth") worms.

The anecic ("out of the earth") worms work deep through the soil profile, often creating burrows which head directly down for over 180cm (6 feet) bringing mineral rich deposits up and drawing organic materials down. They are recognizable, by their large size and darkened heads; these are usually the species favoured by fishermen. The endogeic ("within the earth") rarely come to the surface. They work horizontally, ingesting soil and leaving their castings in the burrows behind them. These species are normally small and have little colouration. As with everything regenerative, diversity is always good.

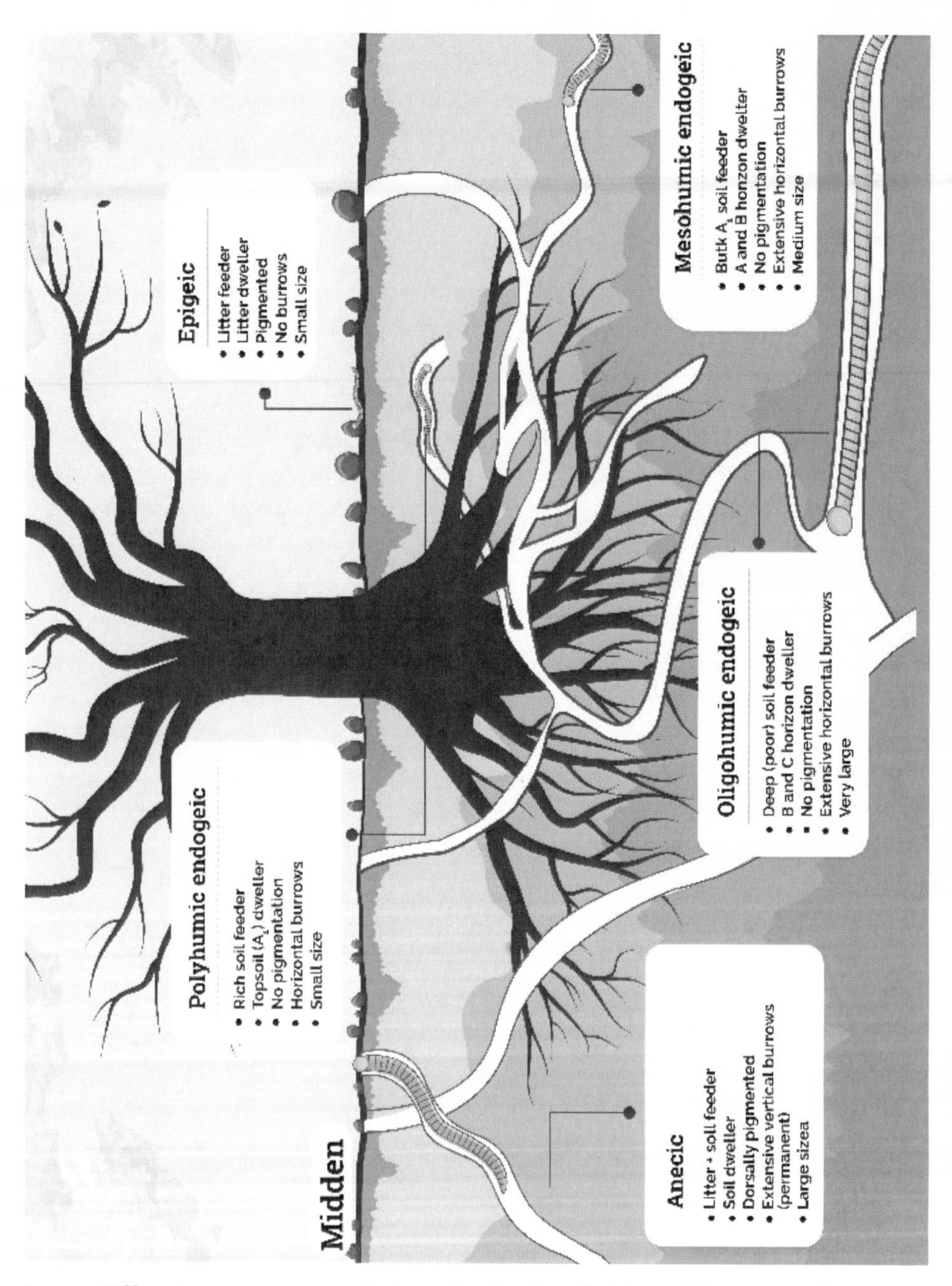

Image: Different worm groups and where they live in soil. Adapted from "How do earthworms affect microfloral and faunal community diversity?" Brown, 1995[67]

Earthworms are not found in deserts or under ice sheets, but it never ceases to amaze me the places we do find them—in dry environments and in ice. Digging holes on a bush-clad farm near Wellington, New Zealand, we came across dozens of ruler-length white worms with bright purple steaks.

Accidentally cutting one with a shovel, I saw that they emit a bioluminescent yellow fluid. I thought we'd discovered a new punk alien species! With some research I found indigenous people had, of course, known of its existence for centuries and had used the worm, *Octochaetus multiporus*, for fishing lures to attract night-feeding fish. While digging soil trenches in Texas with the regenerative grazier and Brahman bull breeder, John Locke, the backhoe revealed apocalyptic scenes as it unearthed and exposed hundreds of writhing worms up to 600mm (2 feet) in length. These worms were impressive, but not quite to the scale of the Australian Giant Gippsland earthworms. I've never had the honour to see, or yes, even hear these giants, as they push their way through the soil. Measuring up to 3 ½ m (12 feet), they're even bigger than the Texas variety.

Many major pastoral worms are European, adapted to a long history of agriculture. Many of these species were accidently introduced into new lands from the soil ballast of ships. Some regions around the world are naturally low in the large, deep dwelling anecic, earthworms, such as the nightcrawlers or dew worms (*Lumbricus terrestris)* and, my personal favourite, the blackhead worm (*Aporrectodea longa*). These major soil engineers have been carried across the continents to provide a source of fishing bait. Others were purposefully introduced in sods of soil to increase pasture production in Tasmania and New Zealand.

In some places, earthworms may be viewed as a pest, particularly in turf industries, with their expensive and carefully manicured sand substrate and in golf greens where worm castings disrupt a perfect putting score. Some of the nastiest residual chemicals like chlordane and endosulfan, were used on turf and school grounds to control earthworms; fortunately, public awareness about the environmental and human health risks of these toxic chemicals led to their bans. In the U.S, chlordane is still manufactured; however, it can only be sold to countries with more lax environmental laws. If you're someone who likes to kill earthworms, you're potentially not the audience for this book. You may, however, know someone who dislikes earthworm activity and the following advice may be helpful. One reason people kill earthworms in lawns, is the lumpy castings they deposit. Often these deposits are concentrated with clay and magnesium, making them hard. In these situations, gypsum and fine limes are great tools. For me, I'd rather remove the "perfect" showcase lawn than poison my earthworms...and my community. In the northern North American forests, which until 11,700 years ago, were under glaciers, the earthworms froze up or moved south. Since the glaciers have retreated faster than a worm can wiggle, the forests have evolved and adapted without the earthworm. It's taken them some time and some human interference, but earthworms are now working their way back North. As they return, they clean up leaf duff and eat and distribute seeds, disrupting modern forest ecosystems. Earthworms are major drivers for plant germination, altering forest dynamics. This has those stuck in the idea of natural restoration, up in arms against the invading worms. I personally believe there is no such thing as "native restoration;" there is no going back in time. We humans have disrupted every ecosystem on the planet, and these systems are now re-establishing new dynamic relationships with what Professor Fred Provenza terms the "new natives."

In hotter and drier environments, it's the ants and termites who fill the niche of earthworms. They are nature's garbage collectors, taking organic materials into their homes, aerating soil and improving water cycles. By increasing water infiltration and reducing erosion, these insects assist producers to increase resilience. Most people overlook these essential benefits, ignoring the insects' feverish scramble as they tread upon their homes. Take a step back and you'll see that plants around the nests grow densely with a more verdant green.

Termites may be key to staving off desertification, creating "hot spots" for biodiversity, moisture and nutrients.[68] As they build their complex tunnel systems, termites stabilise the sides with their mucus and faeces. This material is rich in organic matter and microbes, including the nitrogen fixers. Inside a termite's belly, there exists one of the highest microbial concentrations in nature. These miniature ruminants contain bacteria and protozoa to help the termites break down what others can't: wood, cellulose, lichen and dust. Unlike cows, who burp out their methane, termites fart more than any other organism on the planet, providing a rich food source for the methane loving (methanotrophic) bacteria.

The more species of ants and termites, the better, as they play different roles in an ecosystem. Studies comparing sites with and without, ants in wheat fields, showed a 36% increase in yield, where ants were present due to increases in nitrogen, microbial activity and water infiltration.[69] Many species provide powerful pest control, eating other insects and their eggs.

When the Egyptians first constructed the pyramids at Giza, a group of termites in North-Eastern Brazil also started to build their edifice. These constructions are visible from space and calculations in 2018 show the area covered by their inter-connecting mounds is around the same size as Great Britain! The volume of soil excavated by these tiny insects could build 4,000 pyramids. To date, the queen of this awe-inspiring city has yet to be discovered.

Earthworm benefits

- Improve water absorption and prevent erosion.
- Reduce slaking—increase the water stability of the soil, earthworm castings can take a direct hit by a raindrop and maintain their shape. This reduces erosion and runoff, hence helps the soil absorb water.
- Soil with earthworms dramatically increases infiltration and water holding capacities. A research study in Minnesota cornfields, demonstrated that earthworms increased water absorption 35 times greater than control fields without the earthworms. At 100 nightcrawlers per square meter, 50mm (2") of water could be absorbed by the soil in 12 minutes, versus 12 hours without earthworms.
- If the top meter (3 feet) of soil contains over 25% macropores (earthworm burrows), that soil should be able to absorb at least 220mm (9") of rainfall before running off or ponding.
- Studies involving the introduction of earthworms to pasture soils show an immediate increase in productivity, usually in order of 70%.

Regenerators love to show off their earthworms and dung beetles like badges of honour. Ancient Egyptians revered the dung beetle as the God of creation and rebirth. This love of dung beetles is warranted; these macho soil bulldozers have a direct impact on the quantity and quality of grass, benefiting animal health, performance and beyond.

Digging holes to excavate around and under a cow pat can be hugely revealing as to the numbers and types of dung beetles at work. There are around 6,000 species of dung beetle, grouped by their actions as either tunnelers, dwellers or rollers. The first visual indicators that dung beetles have arrived are the pocked holes in manure; here you may find the dwellers working away with their young.

Globally, the most common type are the tunneling beetles, which show evidence of their work by a ring of mounds of different coloured subsoil around a pile of manure. These tunnelers dig down through the soil profile and carry their precious dung parcels 20 cm (8") deep, to lay one egg into each 'brood ball.' Watching the sped-up footage of dung appearing to boil during a dung beetle feeding frenzy, can be quite transfixing.

Rollers, on the other hand, roll their tasty snacks away from the main pile, to bury with an egg or to eat later. These rollers can pull over a thousand times their body weight; an equivalent human effort would give you the strength to drag a blue whale. As a child, I remember watching David Attenborough gushing over two giant rollers with their heads down, tails up, fighting over elephant dung. Seeing these rollers, still fills me with nostalgia.

Dung beetle adults and the larvae differ in their feeding habits and their mouth parts. The mature adults have filtering mouths, more adapted to sucking up liquids and consuming microbes, while their larvae are the chewers, with mandibles that can grind up the tougher fibres in the dung mass. Adult dung beetles are good parents, often feeding their babies until they reach maturity and fly the nest. Many females only lay between 15 and 30 eggs in their lifetimes. This has an advantage in supporting the success of the next generation, but it also means dung beetle populations can take years to recover from disturbance events.

In every environment where there are animals making poop, there exists a poop-scooping companion, unless they were in New Zealand. When the first Polynesians arrived, there were no mammals, except for a few bats. Instead, the island environment supported a diverse range of birds that fitted into every niche, from small, mice-sized wrens, to herds of elk-like Moa (3m/10 feet tall), to a giant apex predator called the Pouakaia or Haasts Eagle. When Europeans first arrived, the Māori told tales of a giant bird that would dive from the skies and snatch away children. Of course, these stories were dismissed as fairytales, until 1871, when bones were discovered of an eagle measuring 3 m (12 feet) across. Potentially weighing in at 15 kg (33 lbs), the Pouakaia was up to 2 $^{1/2}$ times heavier than a Bald Eagle, with claws as wide as a tiger. Its swooping dive attack has been compared to the impact of a concrete block thrown from an 8-story building.[70] Their prey, the Moa, could weigh as much as 230 kg (510 lbs), requiring a mighty collision to knock them down. When their main food source was wiped out by the Māori in the 1400's, this giant predator lived on only in bedtime horror stories.

It wasn't until the late 18th century, that the ungulates (hooved animals) were first brought across the seas to New Zealand and no one considered inviting their poop-processing partners. The small native dung beetles adapted to bird guano, were literally overwhelmed by the 100 million tons of manure that loads the New Zealand environment every year. The country has one of the tightest border security systems in the world, learning the hard way, from the devastating incursions—both accidental and intentional—of rats, rabbits, possums, wallabies, stoats and spiders. As a result, it has been a protracted process, to reunite New Zealand ungulates with their dung beetle buddies. In 1956, a release of a single Mexican dung beetle species had some success. However, they didn't stray far. The approval of 11 species in 2011, created nationwide excitement, with farmers spending thousands on ice-cream containers of beetles.

After boxes of beetles are purchased, their survival must be ensured. Finding literature about which chemicals are dung beetle safe, can be confusing. Some pesticides kill adults, some only kill juveniles, some reduce egg laying, while most create an environment that dung beetles will avoid. Just because a wormer is 'organic,' do not assume it will always be dung beetle friendly. Natural pyrethrums[71] and garlic for example, may be linked to reductions in dung beetle diversity.

Some anthelmintics (dewormers) are eliminated in the urine, such as Albendazole and Levamisole, with reduced impacts on dung beetles. However, the wellbeing of the wider soil foodweb is potentially still at risk. The pour-ons and long-lasting bolus organophosphates and products such as Abamectin, which kills 100% of dung beetle larvae, are best avoided. Consider too that Ivermectin, when administered as a bolus, can keep killing dung beetles for four months following treatment.[72] Many chemical controls impact upon non-target species directly and indirectly, such as birds, bats and the super-beneficial predatory dung flies. As with many chemicals, there may be delayed reactions and non-lethal effects which are difficult to study. In New Zealand, the spread of dung beetles from their original release sites, is mainly concentrated on those farms avoiding the chemical controls.

Across the world, anthelmintic resistance is on the rise, due to the over-use and misuse of anti-parasitic drugs. According to Nick Sangster, a programme manager for Meat & Livestock Australia, "Nearly all standard anthelmintics have stopped working against sheep parasites."[73] New drugs, multiple combinations and increasing the frequency of deworming, are all short-term measures.

Jules Matthews, a key member of the Integrity Soils coaching team, uses several successful strategies to wean farms off the chemical drench treadmill. Jules has spent decades ranching and farming in America and New Zealand. She completed her Holistic Management training in 1980. She's one of the most knowledgeable grass stewards I've come across. Her ability to increase lamb performance and carrying capacities on diverse, longer rotational pasture systems, is a skill many producers dream of. One strategy to reduce the need for anthelmintics Jules recommends, is to treat the younger and mature stock differently. There is little need to drench older animals. As livestock mature "they develop a natural level of resistance to parasites." It's important to monitor younger stock post-weaning. Without the immunity mother's milk provides, they can be more susceptible to parasites until they develop their own resistance. At times animals can experience more stress, especially towards the end of pregnancy and early calving, which is when they can be more vulnerable to parasites. Support these animals by ensuring they have access to minerals and reduce stress as much as possible around weaning.

Typically, seeing production losses to parasites, points to nutritional problems, stress, or animals with low natural resistance. Some animals may have been bred to perform well in a high-input or feedlot systems. However, if livestock continue to struggle under a low-input system, they need to be removed from your breeding or finishing programmes.

Jules advocates regular monitoring and only drenching if required, to avoid any major stock health issues. Taking faecal egg counts is important. "If you are drenching on visual symptoms of an animal," Jules says, "you have gone way past the point where you should have addressed the issue." If anthelmintics must be used, a regenerative strategy is to leave a refuge for your dung beetles. This can be achieved by not drenching a minimum of 20% of the herd at any one time, so dung beetle populations can continue. During the withholding period, use a sacrifice area to contain animals while the drenches are most active. This will reduce the impacts across your entire landscape. These sacrifice areas can then be placed into a rehabilitation programme, to detoxify the longer-lasting effects on your microbial livestock.

At Prospect Farms in Western Australia, the Haggerty's have not used anthelmintics for nearly 20 years. Their merino sheep are an integral part of the farm, providing weed control and spreading valuable microbes across their lands. The merino breed is a fine-wool sheep, originally from semi-arid Spain. When merino were first brought into Australia, they immediately flourished in the dry climate with nomadic pasturing. They grew tall and rangy (much like Ian Haggerty) and produced nearly double the wool of their southern counterparts.

Ian calls the New Zealand merinos, 'midgets,' due to their compact growth, far better adapted to the cooler mountain living. Centuries of genetic selection in merino have produced the finest quality wool producers, but certain genetic traits have been selected for and selected against, such as hoof quality and mothering abilities. Under Di's steady hand, their sheep have naturally been bred away from many of these drawbacks. The Haggerty's don't mule their sheep. Mulesing involves the surgical removal of strips of skin around a sheep's back end, a common practice in merino wool production. Merino have soft wrinkled skin and wool around the buttocks that can retain faeces and urine, attracting flies. As anyone who has had to deal with flystrike knows, it is truly a tortuous affair for all. With consumers becoming increasingly aware of on-farm practices, mulesing has come under fire.

Is Di being cruel by not mulesing? Just the opposite, they just don't have the fly issues that plague many other local producers. Ever notice how certain cattle in a herd will attract more flies than others, or how a single sheep will run around a field being chased by a swarm? Flies are attracted to the smell of death and decay. These are nutritional problems associated with free amino acids, low sulfur and stress (strikingly similar to the signals emitted by unhealthy plants, attracting plant pests.) Through nutritional feeding and biological soil management, these free amino acids can be complexed into quality protein.

Image: The merino rams at Prospect Farms. Credit Di Haggerty

Livestock carry a huge range of microbes in their guts. When using chemical controls and feeding poor quality feed, these microbial communities are negatively impacted. The Haggerty's have found that by avoiding chemical drenches and using animals raised on their own land, their soil health programme on newly leased grounds, is fast-tracked. Their home-raised animals inoculate lands with beneficial fungi, bacteria and protozoa picked up from the soil treatments. As they have lifted the quality of the crops, stubble build-up is no longer an issue. Looking across fence lines is always insightful; neighbouring paddocks have standing stubble that is grey and dry and sheep listlessly meander, hoping to find a weed to eat. At the Haggerty's, the stalks are bright yellow, and grazing must be managed carefully to ensure some ground cover is left to protect soils from Summer heat and wind. This is the downside of having a system where residues break down too quickly. Fortunately, the long-dormant native grasses are coming back to life; provided Summer rains arrive, the farm will have a living green root year-round.

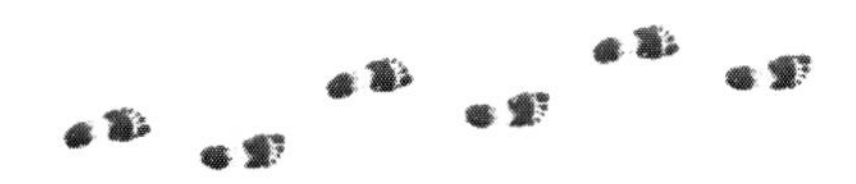

In 2017, Jono Frew, a 30-year-old cropping farmer, left behind a family chemical application business to work at Harts Creek Farm, near Christchurch. Harts Creek was one of New Zealand's earliest certified Organic (1986) farms and today they are leading producers of vegetable seed, fresh vegetables, specialist crops and sheep. Thrown into the deep end, the experience lit Jono up with a passion for regenerative systems. Jono's turning point was a realisation that diversity does not mean plant competition. He had been led to believe that more plant species increases competition and reduces potential crop growth. What he saw at Harts Creek was the opposite; underplanting barley at the two-leaf stage, with ten-way multi-species mixes, outperformed barley monocultures. This shocking realisation, set him up for more experimentation and a deep questioning of the status-quo systems.

During one experiment with lambs, they increased pasture diversity and fed lambs free-choice humate. They followed this grazing rotation with cattle. Just as it is with our own diet, diversifying animal forage produces a diversity of gut microbes, lifting animal health and performance. Many plant species have high tannins, sesquiterpene lactones, phenolic glycosides and other natural anthelmintic properties. Jono has always based drenching on need. Harts Farm takes faecal egg counts to determine the need for intervention. That first year, not a single lamb required a drench. It wasn't a goal or a focus that Jono set out to achieve but, rather, a beneficial side effect from providing plant species with natural worming capabilities and rich mineral sources.

To fast-track your regenerative programme, chemical anthelmintics should be used sparingly or, ideally, avoided. In the transition, supporting livestock with minerals and organic drenches is a good option to support animal wellbeing. I personally advocate the use of the Pat Coleby mix[74] and free choice humates which enhance, rather than destroy dung beetle populations.

At Steve Charter's 2 lazy 2 ranch in Montana, within weeks of the first application of vermicast, fish and molasses, dung beetles flew in to much fanfare. In Steve's living memory, dung beetles have never been seen on the ranch. In areas away from the trial and before 2014, manure could take up to four years to breakdown. Monitoring 5 years later, revealed 9 different species and 700 individual dung eating insects in one cow pat! Depending on moisture levels, manure on the ranch is now breaking down in days to weeks. Within minutes of dung hitting the ground, the clean-up crew swarms in. In this carbon-limited, semi-arid environment where decomposition is slow, dung beetles are providing a valuable tool to the ranch and to the planet. Their action plays a vital role in reducing greenhouse gas emissions; increasing soil carbon storage to depth and reducing methane emissions by as much as 12%. By gobbling up and burying the food resources, dung beetles also leave little behind for pest flies and parasites. This has the added benefit of increasing grazing efficiencies, as avoidance of fouled areas decreases. This helps producers reduce costs and increase livestock numbers.

Natural anthelmintic plants

Plants with natural parasitic properties include forbs such as: Wormseed mustard (*Erysimum cheiranthoides*), Indian pinks (*Spigelia marilandica*), Common purslane (*Portulaca oleracea*), Chicory (*Cichorium intybus*) and Narrow leaf plantain (*Plantago lanceolata*); legumes such as Birdfoot trefoil (*Lotus corniculatus*), Sainfoin (*Onobrychus viciifolia*), Sulla (*Hedysarum coronarium*) and Lotus major (*Lotus pedunculatus*); as well as browse species, like Willow (*Salix spp*), Black walnut (*Juglans nigra*), Oak (*Quercus spp*) and Erect Canary Clover (*Dorycnium rectum*).

Support your above and below-ground livestock

To speed up success, stop killing your clean-up crew!

Cultivation, anthelmintics, carbamate insecticides and fungicides are all detrimental. Avoid using any long-residual chemicals.

If crop residues are slow to breakdown, apply a digestion spray.

Address the six major ingredients required for microbial digestion: air, water, sugar, Ca and a little N, P.

Optimise groundcover, introduce organic materials. Bio-stimulants such as vermiliquid, liquid seaweed/fish, will feed the bacteria and fungi that earthworms and insects love.

If you do use anthelmintics, create a refuge for dung beetles and other insects, by leaving at least 20% of your livestock untreated. Use sacrifice areas during chemical withholding periods.

Reduce animal stress during handling, drought, weaning, calving. Provide shelter. Practice good rotational grazing practices, avoiding overgrazing or set stocking.

Run a diversity of livestock.

Have a diversity of forage; many forbs and fodder shrubs have natural anthelmintics.

Support animals nutritionally, with free choice minerals and humate, supplements and probiotics.

9

High Input Transitions

"If you want small changes in your life, work on your attitude. But if you want big and primary changes, work on your paradigm."

Stephen Covey

We're told fertilisers are simple. When a crop grows, it draws nutrients from the soil, nutrients that are removed when you harvest a crop, sell the milk from your goat, or take steers to the slaughterhouse. At least that's what producers and agronomists have been told, since Liebig first did his NPK plant test. Certainly, there is ample evidence for this in the field. Without fertiliser, many hay producers see yields drop and bare soil increases every year. Dairy farmers produce less milk and ranchers see reduced carrying capacities. It all makes logical sense. Or does it?

Clean calculations bring us a sense of security, something we can control and give people peace of mind. And these calculations do stand true in conventional agriculture. 40 kg of P was removed by crops, so 40 kg must be replaced. An agronomist can predict yield based on the addition of 200kg of N; as long as being outside, doesn't interfere with drought or hail, insects or disease. Many credit NPK fertilisers for the great leaps forward from the Green Revolution last century, but they're not the whole story. These yield responses were due to several factors, including irrigation to new land, new cultivars, machinery and access to credit. Over 70% of rice and wheat, major stable crops, were bred as high yielding varieties, a factor that alone enabled yields to double.[75]

The benefits to producers have been a mixed bag, as these growing methods demand more investment into infrastructure, machinery and land. Over time, input prices rose and the return on products which now flooded markets, dropped. Canadian farmers are no more profitable than they were in 1970 (see graph on next page) and Canada is not unique in this. As a result of declining profit margins, many producers had to "get big or go home", and their children escaped to the cities. In the US, over 73% of smaller rural communities are shrinking, as more people leave than arrive, a pattern mirrored across the developed world. What has been increasing for farmers, is debt. In New Zealand, rural debt topped $60 billion NZD. That's a lot for a country with around 53,000 farms. Debt, long hours and stress is an everyday occurrence, for many working on the land. It breaks my heart to think, that those with their hands in the mud and dust, in the extreme cold, heat and wet, are the ones who receive less on their produce, than the grocery store, baker or packhouse. When you buy a loaf of bread, nearly all your money goes to the middlemen. In the US, this season, a rancher is set to lose money on a steer they raised and finished for 2 years, while the packhouse is profiting $450-500 per animal. Farmers and ranchers can no longer sustain being price takers; to do so, will ring the death knell for those who feed society.

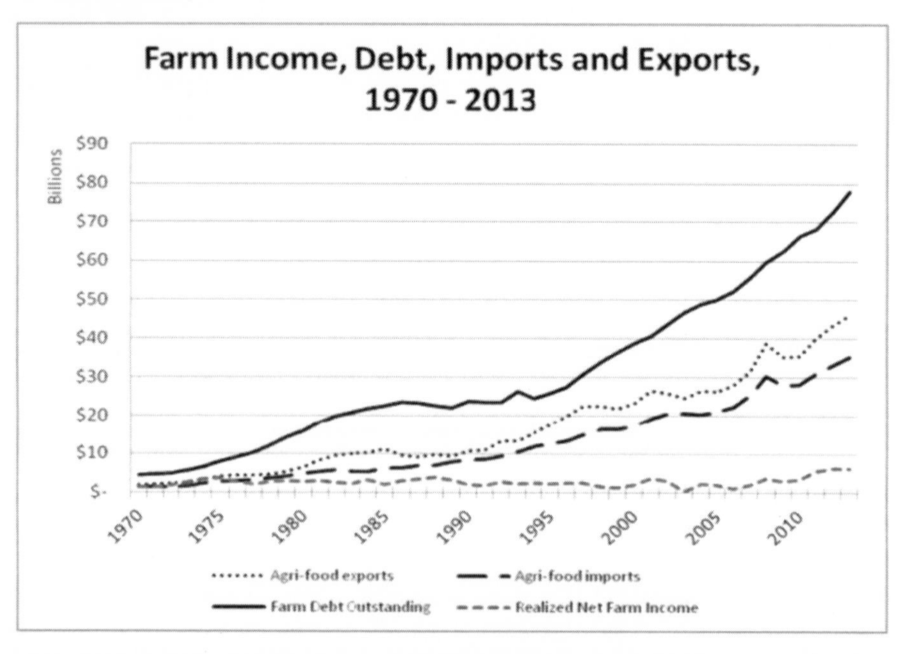

Graph of Canadian farm income, debt, imports and exports1970-2013. Source Statistics Canada.

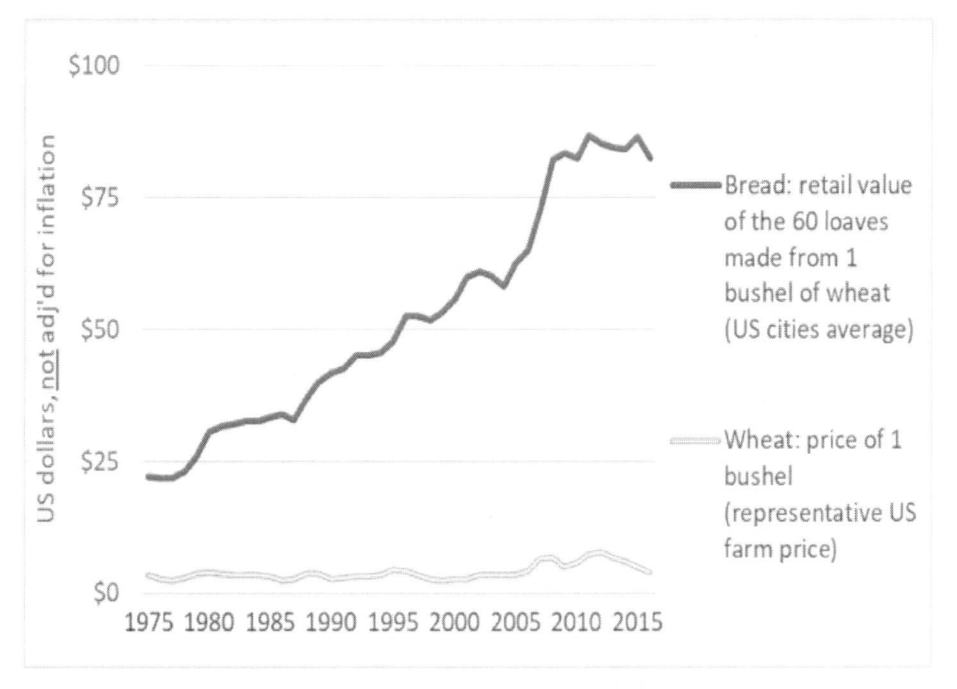

Graph United States retail store bread price and farm-gate wheat price, 1975-2016. Source Darrin Qualmin[76].

If you put together all the different pieces of what it takes to produce food, the Green Revolution has not delivered on its promises to producers. Instead, it delivers to the middlemen, banks, supply and chemical companies. The Food and Agriculture Organisation (FAO), is clear on this: Modern farming practices have increased the risks for food producers, with market volatility and increasing climatic unpredictability. Farming is a risky business, which is interesting for a bunch of people who are widely considered to be risk averse.[77] Nature herself, is a risky and fickle mistress, as all who work on the land know. How to mitigate risk, is the greatest challenge for producers today.

Occasionally, I present soils classes to conventional producers. Preparing for such schools sets me on a turbulent emotional ride, which swings from the exciting potential to shift someone's future, to sheer white-knuckle terror. During one such class in a remote cropping area, the topic was raised: "who would like their kids to want to take over the farm?" The resounding response from the 30 farmers: silence. Then into this void a broad-shouldered farmer slowly stood and from under his hitched down baseball cap, he spoke "Our inputs keep going up. I don't know what's happening with the weather. We have so much debt. I'm so stressed. Why would I want to hand this over to my kids?" It's times like those, that really make me take stock of life and of the difference regenerative ag can have on people's lives. We're not just talking soil; we're talking about a revolution that affects every aspect of rural life. And its time is now.

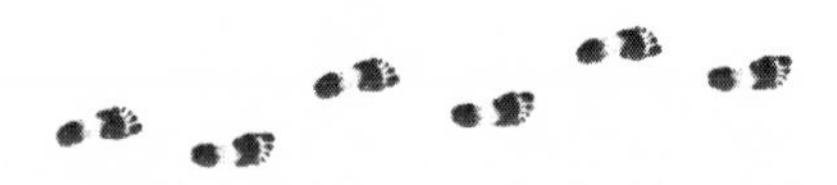

I once was privileged to hear a powerful presentation by soil scientist Dr. Daniel Hillel. In 2012, he was awarded the World Food Prize for developing a method termed "micro-irrigation agriculture" which increases water efficiencies in arid climates. He shared his story of camping with Bedouin in the Arab desert, where he overheard an elder asking his students what 1+1 equals. Their answers were more varied than the stock standard "2", that western children are raised to answer. One child replied thoughtfully, "Well, if it's one nanny goat and one billy, then 1 plus 1 could be 3 or 4.". When working with biological systems, 1+1 rarely equals 2. We often see surprising results, as soil systems function again. As they flocculate, roots penetrate deeper, water and nutrient cycles turn back on and the carbon buffer builds.

There are multiple factors involved in building topsoil, one driver happens from the top down, with biological activity and the other happens bottom-up, through chemical and microbial mineralization. These soil building processes, as I've shown, can dramatically speed up, making previously unavailable 'locked-up', raw mineral materials available to crops. One NZ high country station I've worked with, saw the equivalent lifts of 1500 kg/Ha (1300 lbs/ac) in calcium, in just one year, across treated areas on the farm. That's with no additions of calcium. Dr David Johnson (NMSU), Col Seis, the Haggerty's, Gabe Brown (and many others), are measuring plant-available nutrient increases from 200 to over 1000% higher, just by stimulating the microbial mineralisation processes. This is how soils are meant to function; all without the need for external inputs. Consider, did a fertiliser truck follow the bison around to build the deepest topsoils on the planet?

Every single minute across the planet, 30 football fields of soil are being lost, which equates to approximately 4T of topsoil for every human being - every year! Soil continues to be the biggest export in every food production zone. And most countries are not accounting for it. Concerned? We all need to be with the UN warning there may only be 60 harvests left, before soils become too degraded to grow crops, action needs to be taken today.

To attempt to halt the insidious slide of soil to the sea, agencies around the world advocate planting trees. However, this surface erosion is not due to a lack of trees. Not to knock trees and roots; they're very helpful. However, it's not the trees that hold soils together, it's fungi. When you walk through a healthy forest, with every footstep, you may be treading upon 300 miles of fungal networks. These hyphal networks intertwine and form a sticky web, pulling together the valuable soil crumbs, or aggregates. Fungi are key in forming the larger soil crumbs and increasing aggregate stability. This stability is the capacity for a soil to resist air and water erosion, a key component in the success of regenerative soil systems. In intensively managed agricultural lands, microbial testing reveals these farms are low in active fungi, mycorrhizae and glomalin. On erodible hill country and on the blow away soils, we can halt soil losses immediately, through increasing active fungi and glomalin production.

The formation of well-aggregated soils is the engine room for nitrogen fixation and carbon sequestration. Gas exchange here, is crucial. At the interior of these aggregates, lie anaerobic pockets. An environment without oxygen, is essential for protecting carbon and for the enzymatic process which enables free-living N-fixers to do their job. These aggregates resemble the nodules on a legume and are performing the same job. This is why creating a fine tilth through cultivation is so destructive; undermining soil function, destroying the home for microbes and reducing both nitrogen and carbon cycles. Cultivation can destroy nearly 100% of soil macropores (the breathing spaces over 2mm). This action devastates the biological communities and disrupts water storage between pores, while burning up your valuable soil carbon.[78] Following intensive cultivation, it can take over 3 years for soil structural systems to rebuild.

For cropping farms, encouraging MF can help to overcome the need for starter P-fertiliser applications. In no-till systems, heavy phosphorus fertiliser inputs are not required with intact mycorrhizal networks. There is evidence to suggest, that the presence of mycorrhizal weed hosts, maintains a diverse MF population and promotes highly effective symbiosis with the crop plant. The benefits from MF to maize yield, by maintaining a diverse weed cover crop, outweighed any yield penalty due to competition.[79] Canadian research has shown that encouraging mycorrhizal colonisation, can increase early uptake of phosphorus, improving crop yield potential, without starter. [80]

Now, I'm not saying natural cycles are closed, because they're not. We live in an interconnected world. The global P cycle is driven by organic inputs from animals like birds, bears, bison and wind. Recent discoveries revealed that biomass burning in Southern Africa[81] and dust from the Sahara was delivering phosphate to the Amazon. Phosphate is blowing in at about the same rate it was losing from erosion; around 22,000 T of the stuff every year. In many regions, collapses in biodiversity are leading to catastrophic declines in ecosystem health. New Zealand forests for instance, once dependent upon regular seabird guano, are now hungry for P and diseases are running rampant. Bears in North America were significant contributors of nutrients, including N and P, from their rich salmon diets, apparently yes, they do poo in the woods.[82]

Encouraging biodiversity, brings increased nutrients from outside the farm gate. Insects are a significant nitrogen source to plants (in a healthy system). A recent study in Nature concluded that seabirds are full of crap (at least that's how I interpreted the papers title), with excrement making a global contribution to over 1.3 billion pounds of N and 218 million pounds of P. [83] With birds and insects in rapid decline, their losses have a significant impact on nutrient cycling, which is unaccounted for. A fact which seems to be missing from our agricultural discourse.

Modern practices that create monocultural deserts, are putting the costs back onto farmers, society and the wider environment. Post-modern Regenerative Ag is our opportunity to restore these cycles and put cash back into everyone's pockets.

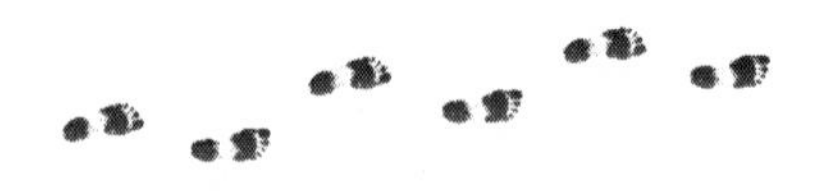

If you drive north in Alberta Canada, past latitude 58, you'll pass a threshold where agriculture lands give way to dense woodlands and waterways populated by moose, beaver and bear. In this land, with perpetual mid-Winter darkness and a sun that just won't set come Summer, you'll discover a surprising pocket of innovative food producers. Tucked between thickets of native aspen, over 30,000 acres of well-managed cropping and cattle lands are being resuscitated back to life.

Jack Stahl, the farm CEO, and his brothers, came north in 1999, to set up Twin Rivers, a new family colony. They are Hutterites, a group who share similar beliefs with the Mennonites and Amish. Hutterites first came to the Americas in the 1870s, escaping Russian persecution for their pacifist beliefs. During WW1, upon refusing the US military draft, four Hutterite men were imprisoned and physically abused. Two of the four men, brothers, Joseph and Michael Hofer, died at Leavenworth Military Prison. It seems intolerance has always dogged these peaceful yet determined people.

Determined, they are, with strong spiritual belief and a patriarchal system of clear accountability. For those raised in the community, the social framework provides a foundation, with clear social expectations that most times, provides more security, clarity and peace, than our modern social structure. Every Hutterite colony has different internal dynamics, depending on the guidance and mindset of their elder males and females. Twin Rivers is a new colony, so cost-effective family labour is not so readily available, as it is for larger, more established colonies. Boys and girls are all expected to contribute to the running of the family enterprises. The boys are in the field or the sheds and the girls carry out household chores, such as cooking, sewing and spotlessly maintaining their beautiful homes. I watched, bemused, as a small, nimble 12-year-old, unblocked a grain hopper, with Jack offering suggestions from the cab of our pickup.

For me, staying with the Twin Rivers family, is an absolute privilege. They are a progressive, outward looking colony, accepting of others, even a kiwi woman consultant. Some of my best conversations around faith and planetary dynamics, have happened sitting with Jack on his porch. He has a perpetual, child-like look of wonder, around what he's seeing with natural systems and biomimicry. He feels they are only just beginning to see the potential of their operation, as they thoughtfully remove the synthetic inputs. His favourite word, said with much relish, is "amaaaazing"!

I met Jack on my first trip to Canada in 2016, after being denied entry to the US. (As an invited speaker receiving speaker's fees, this is not a "grey area", as other officials had led me to believe. A rat-faced homeland security official took much delight in grilling me upon the finer points. "This is work and you do not have a work visa!") Spending an unplanned two months in

Alberta, Canada, with the passionate Graham Finn and his family, led to many unexpected new encounters. Graham, with his company Union Forage, is an example of how globally connected we all are; here's an Australian, selling New Zealand and European seed to Canadians. He is a wealth of information about cover crops. The world is indeed a small place. An unexpected outcome from my hiatus in Canada, was proudly returning to the US with a shiny new visa for "An Alien of Extraordinary Ability"! The other positive outcome, was meeting Jack and Josh.

I was in Canada to teach a class with the Foothills Forage and Grazing Association, a remarkable producer driven extension group. They focus on providing training and demonstrations around profitable and regenerative production methods. Jack and his brother Josh, the farm boss, came to the soils class. Josh sat in the front row, arms folded, in his traditional suspenders, unwilling to suspend any disbelief at what Jack was leading him into. Josh had done his homework and brought a list of prepared critical questions.

To say he was skeptical, is putting it mildly. The family overall, is a wary, thoughtful group. They do not open their doors to many. Jack says he'd done a good 4 years of research, before they decided I might be safe to invite up to the farm. My first day with them, felt like I was back in high school; being drilled for final exams, first by the CEO, then the farm boss, the cattle boss, the garden boss, then all the nephews and sons. Jack announced he was going to "Out-Gabe Gabe" and laid out his dream to build soil carbon faster than the legendary North Dakota farmer Gabe Brown.

When we met, Twin Rivers was already in the top 1% of producers in their region. They were coming off two of the best production years the farm had ever had, making it all the more remarkable that they chose to do a total U-turn on a farm that, conventionally speaking, was working perfectly. "We were starting to hit brick walls in the poison system," says Jack. He, like many others in this book, saw profit margins shrinking, with increased needs for more assets and inputs. He doesn't mince words: "So many farmers are laundering money for the chemical and equipment dealers. They are living poor and dying rich."

Image: Harvest time at Twin Rivers 2017

Jack is a visionary, who prides himself in his ability to plant seeds in people's minds, which he enjoys watering. On my very first meeting with the Twin Rivers team, Jack brought me into the coffee room, the control room for Hutterite operations. An intimidating crew of around 30 men of varying ages; they all sat with their checkered shirts and hats and a healthy air of distance. Jack opened the meeting with, "Tell them what you told me." No pressure Jack! Thus, I shared with them stories of transition and possibility and what others are seeing around the world. And so, our journey together began.

The Stahls' know they can make good even better. Their careful observations of their land led them to conclude, that their current success was unsustainable for the long haul—the ultimate test for a family-based operation that gives weight to legacy in their decision making. They watched as their soils progressively declined, becoming increasingly tight, with precious rains ponding on the surface. Digging holes was hard labour, revealing stunted and sideways growing roots. Since 2016, they've been growing around 1,000 acres of cover crops every year, when germinating rains are not limiting. These covers include a diverse mix of sunflowers, vetch, triticale, oats, kale, rape and radish. Even the powerful sunflower and brassica root systems struggled to make headway in the hard grey-wooded clays.

The aspen woodlands, the native groundcover here, created the black and grey-wooded clay soils, with low natural carbon levels. These soils are powerpacks of major cations, Ca, Mg, K, but sodium (Na), was non-existent. As a result, from compaction and poor biological activity, their trace element mobility was poor. Our first limiting factor to address, was how to get air into the system. Working with a local humate company, we came up with a unique recipe for a prill with gypsum, humates, sodium and trace elements. A prill is a granular product; with the humate, it creates a powerful, efficient, slow-release fertiliser. Around the world, companies have been using this technology for decades, to buffer fertilisers with carbon-based carriers such as lignosulfonate, compost and humates. By placing trace elements and other nutrients with the humate, biology can effectively release nutrients when plants need them and at the correct rates.

Jack doesn't believe in trials. He said, "I trust you," and immediately put 22,000 acres onto the prill programme. "If you've got 2 feet in the fire," he likes to say, "you don't pull one out to see if things get better!" He chooses his advice from people he believes have already done the hard work before him like Ray Archuleta, Walt Davis, Dr. Kris Nichols and Gabe Brown.

At that point, I'd never worked in Canada. I didn't know local products. But Josh trusted me and the principles of soil health, so off we went in with no plan B. After 3 days of blocked fertiliser augers and machinery, Josh started questioning my rationale. The prill had come out too soft and was bridging, so he added some canola oil to make it flow and the 2017 season began. They planted a range of cash crops including canola, wheat, barley and peas across multiple soil types spanning 25 miles.

It's always better to start a biological programme with some moisture, to fire-up the microbes and kick-start the carbon cycle. After the snows melted and the soils dried enough to get the seeder out, the rains stopped. That entire season, the farm received a patchy 127-180mm (5-7 inches) of rain. Cover crops sown later, barely germinated to cover the ground. Within a few weeks, Josh was getting nervous. His neighbours crops were up with no sign of his. He is a vigilant scout and was encouraged to dig a hole and see what was happening. He was surprised to find, well-structured root systems deep in the ground, already deeper than what the previous crops had been at the end of last season. This is not uncommon with regenerative programmes. Having plants primed with good root systems from the start, sets a crop up for success. Although the crops might take longer to start above-ground, they quickly catch up.

Our first-year programme, immediately dropped nitrogen inputs by 75%, from 123 to 33kg /Ha (110# to 30#/Ac). Josh was suitably nervous about such a considerable drop. Most days, I was quietly confident we could do this; our leaf tissue tests showed that N wasn't a limiting factor and in places was excessive. Nitrogen uptake to the plant is affected by compaction and excess magnesium.

There are key bacterial groups critical to the proper function of the soil nitrogen cycle. These bacterial groups work in a chain, with each group making a nitrogen 'product' and passing it along to the next member of the chain for further processing, ultimately into nitrate. If any part of this chain is weakened, the entire nitrogen cycle can effectively grind to a stop, reducing the amount of N available to crops. These bacteria hold N in their bodies. Over 80% of the N entering the plant, is liberated by grazing protists and nematodes, mycorrhizae and viruses. There are also free-living N-fixers in and on plant roots. Applying high rates of N (which depends on soil types but aim for less is more) halts natural N-fixation processes. This breakdown in the N cycle, can occur due to low biological mass, hydrophobic soils, waterlogging, soluble N, compaction or under/overgrazing. Stop to consider, before you reach for a bag of urea, the issue may not be nitrogen; it may be, you're missing the predators.

Ideally, for soil structure in heavy clay soil, we like to see Mg around 12% Base Saturation (versus 20% in lighter, sandier soils). At Twin Rivers, Mg ranged from 30 to 41%, well above excess. This high Mg pulls soils tightly together, creating hard, poor infiltrating soils. By putting gypsum and humates down the slot, we flocculate and flush excess Mg around the plant root zone, reducing the need for N inputs. The gypsum ($CaSO_4$) reacts on the colloid producing soluble $MgSO_4$ (Epsom salts), swapping out excess Mg with the Ca. This reaction doesn't change soil pH. We also added humates to our nitrogen, which automatically increases N efficiency by 30% and feeds our beneficial fungi. By increasing our effective N use, the difference in actual N to the plant, is not as dramatic as the numbers indicate. If you are currently using a lot of N, adding humates is a must, this one tool can help you reduce costs by 30%, without any risk.

When talking about applications of nutrients, phosphate is always raised. "You can't continue to produce crops, without phosphate fertiliser" is the catch cry for conventional chemical agronomists. Phosphorus drives the energy system of all cells. It's the 'P' found in adenosine triphosphate (ATP).

ATP is the "molecular unit of currency" that runs the photosynthetic engine on your farm or ranch. Without adequate P, you will not have nutrient density or optimal Brix. You need P for good germination, root penetration and fluid transfer within the plant. Plants with poor P mobility grow stunted. If you see them with a slight purple/red tinge, P is already critically low. The purple colour comes from anthocyanins, which plants use to protect themselves against sunburn. It's the stuff that gives red wine its colour. Anthocyanins are antioxidants, not that you want to stress your plants to get a response, however. There are large areas across North America, where you can see this purpling across the range. In livestock, phosphate deficiencies show up as poor milk production, growth and cycling, downer cows, stiff joints and lameness.

Modern farming systems struggle to deliver P efficiently to plants, with close to 90% of P becoming 'locked up' or lost through soil movement. Phosphorus is infamous, for its tendency to tightly bind to cations Al, Ca and Fe. All soils with a history of applied fertiliser, contain large amounts of insoluble phosphorus that is locked up or "fixed" in soil. By using a chemical farming approach, it stays relatively unavailable. Another unintended consequence of soil surface nutrition, is that shallow feeding roots are encouraged, leaving plants more vulnerable to climatic stress.

Excess phosphorus has long been recognised, as a leading cause of pollution in waterways, but many don't realise the negative impacts on food quality too. There are also theories that excess P leads to cancers. Studies in rats show high levels of skin and lung cancers. New Zealand has the highest cancer rates in the world, despite its natural environment and clean air. NZ also has the highest P fertiliser use/Ha and highest P levels in foods. What does phosphorus do again? Oh, yes, cell replication.

Across many rural areas, there are operations which have been applying raw manure from feedlots over long periods of time, leading to an excess of many salts and minerals including P. I once visited an 'organic' dairy farm who were applying manure for 3 months every year, using 3 trucks, running trips to the field all day. They loved their manure! Instead of P levels on a plant available Morgan test, being 170 ppm, theirs was 1700! Their Bray P test was 294ppm, when 24 is ideal! High P can cause drops in milk production too, with poor thrift and scours in calves. Their grass levels were terribly imbalanced, yet their cows looked fantastic. It took all day for me to figure it out. "Oh yes" said the farmer, "we feed them free choice seaweed. They eat about $100,000 of it a year"!

If you want to reduce your P input, it's important to understand a little of its dynamics in plants and soil. Nearly all phosphorus availability is governed by organic materials, plant roots and biology. As phosphorus is made available primarily by micro-organisms (75% and greater), soil mineral tests may not be the best indicator to assess P availability. Optimising root development, using diverse cover crops and stimulating phosphorus solubilisation with bacteria and fungi, is key in mobilising P. There are specific species which solubilize phosphorus, including *Azotobacter, Pantoea, Microbacterium, Pseudomonas* and *Baccilus* species. Mycorrhizae play essential roles in plant nutrition, as evidenced by their beneficial effects on pasture growth, phosphorus nutrition and water relations.

As a general rule, we can track this mobilisation using plant tissue testing and see that even with reduced volumes of P inputs in a biological programme, leaf P levels keep lifting. At Twin Rivers, we take plant tissue tests through the growing season to track plant health and pick up any early warning signs. The 2017/18 tests showed ideal phosphorus levels. A total of 1.1kg of P was applied all season. At sowing, a seed dressing was used that included three specialist bacteria, *Rhizobia, Azotobacter* and *Bacillus subtilis*, plus AMF were used. When using inoculums, there are synergistic benefits when applying diverse blends of microbes.[84] These bacteria provide faster release of rock phosphorus and enhance phosphorus release to your crop, throughout the growing season. They have other added benefits for plant resilience, rooting hormones (auxins), water accessibility and root protection.

Treating seed with microbial extracts, is a powerful tool for plant health. As Dr Johnsons work with the bioreactor shows, seed treatment results in increases in yield and plant health. Other research has shown this practice increases seed germination, crop nutrition and resistance to stress.[85 86] Plants and microbes have complex interconnected relationships. Research is discovering that plants employ an ingenious method to feed their micro-herd and then reabsorb their bodies in a process called the "rhizophagy cycle." By reabsorbing these microbes, plants can then access bioavailable nutrients. Researchers estimate that potentially 30% of nutrients taken up by seedlings, is derived from the bodies of these sucked up microbes. [87] Dressing seeds with their microbial partners makes a lot of sense in transition.

Across our monitoring sites (except for boron in the peas), the trace and major elements all came up to desired levels on tissue tests. Josh regularly sent in photos of beautiful even crops, peas covered in pods and well-developed root systems. Their crop harvest was comparable to the neighbours, but at substantially lower cost. Unfortunately, the 2017 wheat protein levels were disappointing, as it was for other producers in the region. In both the 2nd and 3rd year in transition, the start to the seasons began slowly, with prolonged dry spells. Only 1.75 inches of rain was received in the first 2 months of growth in 2019. Three years of poor climatic growing have been frustrating. These are not the most ideal times to try and put our best biological foot forward. However, by starting in poor seasons, their success cannot be accounted to good rainfall and climate.

Twin Rivers 2017 Soil Program			
Product	**Application**	**Rate/ac**	**Notes**
AMF Trichoderma P-solubilising bacteria	Seed treatment	300mls	Applied per Tonne of seed
10 10 10 NPK Fulvic acid Boron (21%)	Foliar On peas only	14 litres 300 mls 0.2 pound	Total kg =0.6 N, 4.7 P, 3.4 K
Soluble humic	Furrow	1 litre	
Gypsum Humate Sea minerals Boron Zinc Copper	Furrow	35 pounds 25 pounds 4 pounds ½ pound ½ pound ½ pound	This is a commercially tailor-made product. In year 2 the gypsum was applied separately. Recorded as actual amount of trace elements
Urea	Between rows	25 pounds	

Chart of Year one inputs at Twin Rivers. This is not a general prescription; this is a programme to address specific enabling factors on this farm.

Aside from the mindset changes, one of the biggest shifts to date on the farm, is the profit margins. Most conventional agronomists will tell you that it's too risky to shift to a low-input system. They use fear as a tool. With narrow profit margins and rising debt, most producers become paralyzed and unable to experiment. In the first year, Twin Rivers saved $100/ac in inputs. In 2018, their break-even was one-seventh that of other producers in the area. Jack thinks they win no matter what, "even with grain prices, climate, or how many bushels of crop you produce, we're ahead." There are many eyes scrutinizing the Twin Rivers operation, despite their desire to remain under the radar (sorry, chaps).

Jack, like many of my North American contacts, is a graduate of the Ranching for Profit (RFP) school. Every day, he reflects upon a different aspect of what he learned at the school. "I'm certainly hugely grateful to Dave Pratt at RFP. He was the one who opened the door for me into the US and the incredible opportunities I've had here." Dave's school is a powerful, week-long transformative process, that leaves people with new accesses to action. Actions that would not have been feasible or even conceivable before the school. Jack is clear, "a change is coming and needs to come, or all of agriculture will be in deep trouble." He is concerned for a farming community that finds change challenging.

"If a ranch or farm isn't profitable, then it's not a business, it's a hobby," Dave says. "Most ranches are a big collection of expensive assets and low-paying, physically-demanding jobs." Apart from shifting paradigms, RFP teaches three simple solutions to increasing profit: cut overhead costs, improve gross margin per unit and increase turnover. Most producers I meet, don't understand what gross margin per unit for their operations are or what's been putting a drag on their profitability. Twin Rivers is not only putting more money in the bank, it's paying them more dividends into the future, through building up their principal bank account, their soil.

Regenerative agriculture is about harnessing life and life is all about carbon. Carbon is the planet's living currency, what microbes use to trade with plants for nutrients and the building blocks for all life. Humates are an excellent carbon tool to support the transition of high input properties. Raw humates are concentrated sunlight energy, sourced from plant materials, peat, soil, composts and soft coals, like leonardite and lignite. The range of humic substances on the market, is as baffling as the range of claims and information about which products are best. There are over 200 humic products just in Australia! I'm in the camp that wants a concentrated carbon, that contains as much quorum-sensing molecules as possible. Worm extracts and soft brown coals fall into this category.

Any time a fertiliser (or herbicide) is required, we recommend a carbon. Humic or fulvic acid are easy to source and to use, and if available, vermicast extracts are ideal. As a general rule, think 'fulvic for foliars,' and 'humics for humus.'

This multi-use carbon is receiving a huge amount of agricultural scientific interest, due to its benefits for plant nutrition and mineral chelation. Chelation is a process which allows a nutrient to "maintain its own identity" within the spray tank and prevents nutrients from being tied up with other nutrients or chemicals in the tank. Studies have shown that humates (which include fulvic, humic and compost extracts) are highly mobile in the plant. They increase the efficacy of foliar applications, through increasing the permeability of the leaf cell walls.

Humates and other biological foods, also help to stimulate the beneficial microbes that live on plant surfaces and in the soil. These organisms are essential in recycling nutrients, increasing fertiliser efficiencies and reducing fertiliser losses, while converting soil nutrients into more plant available forms. New Zealand research on sheep and beef stations, by independent scientist, Dr. Peter Espie, has led him to conclude that the "biological enhancement of plant growth and nutrient content is scientifically valid." Other studies on biologically managed dairy farms in Rotorua, New Zealand, have shown that these farms had significantly lower nitrate concentrations than conventional farms. Only small amounts of humic products are needed. Humic substances are a concentrated battery of microbial foods. At the soil table, microbes feed before plants. Adding high volumes of soluble humic products will put a drag on yields. Remember to just "tickle the system."

Although solid chemical fertilisers may contain higher nutrient concentrations compared to foliar fertilisers, they are highly inefficient and little of this is actually available to the plant. Fertiliser use can be maximised through direct application to leaves and roots, enhancing the uptake and utilisation of plant nutrients. This method bypasses the biological gut system, enabling producers to overcome temporal imbalances during a transition. With the addition of a bio-stimulant, you can keep your soil health goals moving in the right direction.

Foliar feeding took off in the 1950s, when Michigan scientists discovered that plants could take up nutrients through their leaves. Since that time, the use of foliars has been hotly debated, as results can vary with different factors. These factors include; soil type, P-retention, pH, soil calcium levels, nutrient diffusion and the time that dissolved nutrients remain on the leaf. There seems little doubt that where soil fixation exists, foliar applications of certain nutrients, is the most efficient method of fertiliser "placement," especially during pinch periods when nutrient demands may outstrip a plant's ability to supply itself. Phosphate which remains on soil surfaces is more prone to losses from soil movement, resulting in water quality issues. Fluidised applications of nutrients penetrate the soil surface to deeper levels, reducing the risk of losses to waterways.

Foliar fertilisers can certainly result in net profitability to producers, when applied in a timely manner, providing a short-term solution to many nutrient deficiencies. With advancements in knowledge and research around chelators and inoculants, foliar nutrition can be implemented to stimulate biology as well. It is now technically possible to increase the efficiency of fertilisers. Take your advice from plant tissue tests and respond using grams or ounces of product. There is no need to hammer in trace elements and minerals.

2019 was an extremely climatically challenging season for Twin Rivers, when early rains didn't arrive, the decision was made to not spend money on the recommended foliar application. In transitions, foliars can be invaluable fast-tracking tools to ensure crop yields are maintained. By not investing in a foliar, the wheat yields were depressed, while the barley and pea crops produced comparable yields to neighbouring farms; fortunately, at less cost per acre. The team is keeping their eye on their long-term goals, building soil health while balancing their profits.

Before they started with a regenerative programme, Jack felt the property was only reaching 20% of its potential, with low functional humus and poor water infiltration. "The place just keeps getting better," Jack says, now with the goal to increase cattle numbers fourfold. If they can store the Spring moisture, they can drought-proof their farm. This 2019 season provided breakthrough epiphanies for the Twin Rivers team. By the end of July, with minimal moisture, germination across the region was sporadic and crops were struggling. Nearly 2 million acres across Alberta were ablaze in the mass of tinder dry fuel. In a 26-hour period, 5 inches of rain bucketed down, putting out the last of the fires and drowning the Manning area. Driving the long miles to sample across the property, ditches alongside their fields were empty, while neighbouring properties ran full and fast. With Josh and Dwayne, a teen soil advocate from the farm, we walked deep into neighbouring fields; an inch of water lay across the surface and soils were impossible to dig, as they were immediately inundated in water. Thirty feet away into the Twin Rivers fields the water had soaked in and the soils were fluffed up with biological activity sent into hyper-drive. Ever staunch Josh, was overcome with emotions, walking through his fields shaking his head and muttering "wow, O wow!" The theoretical possibilities for Josh, have well and truly become a living and breathing reality. In these landscapes and indeed across all agricultural areas, if we can capture and hold onto every drop, we are set for success. Not only does the farmer win by catching their water, townships downstream are buffered from flash-flooding events.

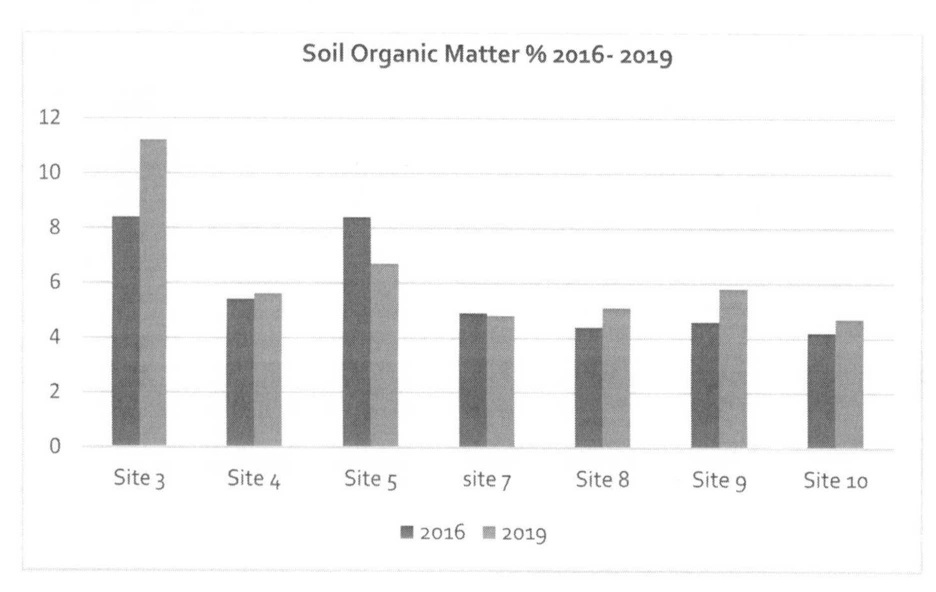

Graph showing changes in OM from 2016 to 2019. Despite the poor growing seasons, over a 3-year period, most of the benchmarking sites showed significant lifts in organic matter. The only 2 sites with a decline in OM, were swath grazed as the snows melted, causing a significant amount of soil damage. Site 3 had large visible changes in soil structure, moving from compacted states to a light, fluffy and aerated soil.

Another noticeable change on the farm is the contagious excitement from the next generation of Stahls; the nephews and sons, are now educating themselves and are driving decisions around biological inputs and soil management. I sat with two teens to teach them the basic use of a microscope, when I looked up from a slide, the room had filled with a dozen young men, avidly listening and helping to write words on the board like "actinobacteria" and "nematode." Until an adult entered the room and

When transitioning, avoid going cold turkey on a high input system, unless you address the enabling factors immediately.

If you can afford to absorb losses by spreading risk across your whole enterprise, then drop chemical use in one area at a time.

Stop the actions that are causing the harm! The pesticides and fungicides are part of the problem. Focus instead on nutrition and stimulating the microbial workforce.

Address all 5 M's: Mindset, Management, Microbes, Minerals and Organic Matter. It may only take great management and a small amount of trace elements to kick-start a lethargic system.

Stop muddling about! It'll take you decades to lift performance if you dabble here and there without a plan. How many seasons do you have left to get it right?

See the Appendix for more Transition Tips.

barked: "there's work to be done outside!" and they scattered like mice when a kitchen light is turned on.

During the soil health journey, both Jack and Josh have reflected on the noticeable reduction in stress, for everyone. The Josh I first met in 2016, is not the man I know today. He's truly caught the soil bug and has a new enthusiasm for life, despite the original equipment blockages. It's now with glee that Josh says, "no plan B!"

10

Measuring Success

"We often forget that we are nature. Nature is not something separate from us. So when we say that we have lost our connection to nature. We've lost our connection to ourselves." — Andy Goldsworthy

Fraggle Rock is an 80's TV show featuring Jim Henson's colourful puppets, covering topics like waste, the environment, spirituality and dealing with social conflict. It's a silly show that resonates with me. The Fraggles family includes a character called Uncle Traveling Matt. Matt has a grey 'Doc Holliday' mustache and with a pack on his back, he leaves his safe underground community, to record his observations about the human's world, (outer space). His backpack is full of measuring tools and notebooks to enthusiastically scribble his thoughts.

In a nutshell, I've just described to you one of my favourite ranchers,[8] Roger Indreland. I think it may have been a shock to this well-spoken and educated rancher, that he reminded me of an 80's TV muppet! It always brings a smile to my face, when we don our backpacks and set out to check on our monitoring sites, under the blazing Montana sun. We've been doing this since May of 2014.

Roger and Betsy Indreland had doubts when we first met at the Ranching for Profit (RFP) Summer conference. Roger's mind has a carefree, child-like quality, the suspense of an unopened parcel or working out how a toy is put

---

[8] A quick note, all the ranchers I work with are my favourites and as I'm currently parked up in my horse trailer on their front lawn, I'm eternally in the family's debt!

together, can drive him to distraction. This need to understand things deeper, left him intrigued enough to confront me to comment: "We couldn't make that work," I flippantly commented, "Well, why not?" They were both baffled that I could be so confident ("cocky" is the New Zealand term), so we set to work. Roger and Betsy have curious minds, but they're not ones to take unnecessary risks. They've learned over the years to observe and record changes, before rolling actions out over their 7,000-acre operation. With their daughters Kate and Ann, they built a successful registered Angus stud, via trial and error---results they are happy to share with the neighbours. They're a family that have earned the respect and admiration of their community, through their contribution to strengthening the Angus breed and commitment to improving the resilience of ranching families. When Roger talks, people listen.

Betsy is his perfect counterbalance, "she speaks louder with her body," jokes Roger. With Roger, the visionary, Betsy carries the details in the pockets of her keen mind. He'll rattle off a sentence, which Betsy completes; "it was cow 5409 that got bit by a rattlesnake last year," "no, it was 5475." Roger knows she's right. They are a powerful, collaborative team, who value their currency of communication.

To stand in the rolling pastures here, you get a sense of what gives Montana its 'Big Sky' reputation. When the sun starts to rise, the light catches the Absaorkee Beartooth Mountains to the South and the Crazy Mountains to the West. Less than 2 hours to Yellowstone National Park, you could be forgiven for thinking you'd found Nirvana—until the Californian fires smoke out the mountains, and the skies turn Hades orange.

At the beginning of the 19th century, the famed Lewis and Clark Expedition, were the first Europeans to cross the western part of the United States. They set to survey the geography, plants and animals as well as to establish trade with the First Nation peoples. Their voyage took them within 12 miles of the Indreland Ranch where Clark noted "Buffalo is getting much more plenty."[88] This area was once dominated by dense, short prairie grasses. Now the rangeland is dominated by sagebrush with bare ground, cryptogams and diverse native flowering species. With early overgrazing of sheep and cattle, estimates point to a decline in soil carbon between 30-60% since Lewis and Clark first sailed by.

Anyone who lives and works on the land, appreciates the forces of nature and how small and powerless we really are. The climate here in 'Big Timber' is not easy, with Winter bringing whistling winds and dense snow, followed by scorching Summer sun. The ranch sits around 4,500 feet in elevation, with 300 to 350 mm (12 to 14") of annual precipitation. The frost-free window is only 90 to 100 days. The bulk of feed needs to be grown in a short and frantic 30-45 days, to supply enough grass for the coming year.

When Roger was growing up, his father was adamant that fertiliser was expensive and made the ground hard. Of course, after heading off to Ag School to study economics, Roger came back with the inflated opinion, that his dad was backwards in his thinking, and that "bigger is better." The climate's irregular rainfall meant the fertiliser numbers didn't really stack up. Instead, they invested in equipment for farming and haying. A dry year hit early on and with repayments due and no harvest, the early years for the new couple were tough.

During college, Roger had the opportunity to work with one of the icons of the registered Angus business. During a private conversation, the breeder revealed a pivotal insight, his belief "the Angus breed was now beyond its optimum size." With Betsy's background in marketing and Roger's keen observations, they knew going to bat against the big Angus players and wealthy landowners, was a risky maneuver, given their promotional budget. The Indrelands were early pioneers in using the genetics from what has now become one of America's most popular "bigger is better" bulls. This decision cost the pair dearly, with 75% of the cows returning infertile, under the tough nutritional conditions in their windswept land. Roger and Betsy have always valued diversity and that year chose to use two different bull genetics. Fortunately, their second-choice bull had much better cow energy value ($EN) and his daughters thrived. Their ultimate realisation? These bigger-framed, high-input breeds didn't have the traits necessary to survive and thrive in an extreme low-input natural system.

These two early incidents were catalysts for the duo to look at techniques to reduce inputs and produce a fit-for-purpose herd, that can perform in low energy environments. The Indrelands have developed their herd by emphasising traits that enable them to perform well in tough conditions, requiring minimal handling or supplemental feeding. For instance, in the harsh 2017/18 Winter, before the snows became too deep for them to dig through, their cows were only fed for a total of 3 weeks. Roger is confident to impart, "We have a herd of cattle that are low input, extremely durable and very sound." This trouble-free line of cattle that works in sync with nature, has attracted a loyal and growing client base interested in resilient, cost-effective and profitable progeny.

Having an indicator for how much input an animal requires to grow and produce milk, is a valuable decision-making tool for producers interested in profitability, rather than showy large frames. Cow energy values ($EN) are used by breeders to predict how much a cow's energy requirements could save you in feed costs. $EN is expressed in dollar savings per cow, so a higher value is better. Not all cultivars or livestock selections are designed for low-input systems. Being able to calculate potential costs would be invaluable in any sector--for wheat, apples, vegetables...but maybe not horses?! In Montana's extreme environment a low, negative $EN means high amounts of supplementary feeding will be required through Winter. If you look in breeder catalogues, the average Angus breed has an $EN of minus $4.01. The average for Indreland bulls this year; +$20.57. Now that's Montana tough.

Until 2006, the ranch was running an approach typical to the area, removing every blade of grass, feeding hay for a large part of the year and then calving in Winter to produce larger calves at weaning. Calving in Winter is a stressful approach for anyone, with lots of sleepless nights and long days ensuring calf survival and cows are well fed. It's a common practice, often in the most inhospitable environments, with warm wet calves hitting frozen surfaces at birth. Seeing cows with no ears and tails can be a clue as to the climatic conditions on their birthday.

After attending the RFP school, the pair had a bombshell moment; they had been calving at the exact opposite time for cow nutritional needs. They shifted calving from February to May, to match the Spring growth and every man, woman and beast breathed a sigh of relief. This timing is a closer match for when wild antelope and deer are birthing on the range too. After my first visit with the Indrelands, we discussed the concept, that these lands may never have been grazed every year by large herds. This thinking was an epiphany; in brittle landscapes, grazing may not have happened for 3 years. Extending this recovery time has been a breakthrough and a return to what Roger believes used to happen before the invention of large round bailing equipment. For drought resilience, "successful old-timers knew they needed to have at least 18 months of feed in front of them," he says. Drought in Montana is not a someday/one-day concept, yet memories can be very short in every farming/ranching community.

Grazing is recorded in a notebook and a chart on the wall. They aim to graze different pastures, at different times of the year. Some of the larger pastures create management challenges; these areas are split up with electric fence. Plant species on the range provide diverse pickings. The irrigated grounds and lowland areas consist of smaller pastures where stock can be moved more regularly. The ranch practices the 'Bud Williams Stockmanship' approach to low-stress animal handling, creating a relaxed environment for people and animals. Most cattle work is done on foot and with their stock dogs, Lily and Ace. Another consequence of the smaller framed livestock is the safer conditions in the yards. Betsy recalls how, in the early days, they couldn't see over the backs of cows, which made her feel vulnerable in small spaces. The cattle "are responsive to us," says Roger, "and that is a source of great pride for us too, to be able to go out and maneuver cattle and do just about anything we want, without any huge problems."

I fancy myself as pretty sensitive when moving stock. However, after watching Roger on foot, silently pull a cow and calf from the herd and wind them up a hill through an open gate, all without breaking a sweat, I am left with a new appreciation for the art of moving cattle.

There was no crisis that led the Indrelands to shift their practices. Roger had a good early foundation with an observant, patient and skilled stockman father. He doesn't worry about what the neighbours think; in fact, I think he enjoys making them scratch their heads. He's unusual in that he's not afraid to ask any question, or anyone. Their neighbour, Gretel Ehrlich, poet and author of the "*The Solace of Open Spaces*," has an ethereal way of capturing the raw nature of the ranching community. She is astounded by Roger's mind. "He stays contemporary in his thinking processes," she says. "He's stimulated by looking at the land." Even when Roger is knowledgeable about a topic, he will still ask people deeper questions to expand or question his own knowing. Gretel and he, both ask the big questions many are afraid to ask. This is a rare skill; indeed, their fearlessness is inspiring.

"What are you doing here? Because your father said you should be a rancher?" reflects Gretel "because you love it? To make money?" Or to leave a legacy? "Most people are terrified to stop" she muses, "you could ask this about your marriage, children or life." Gretel is a close friend of Allan Savory and has travelled and visited with him on many ranching operations. She marvels at the same process many discover when deepening their relationship to soil. "It's the new thinking which makes it fun, which no one ever anticipates." The new learning and creative actions dissolves "the poison of tradition," she says. She's a lyrical genius. I'm in awe.

Coming from traditional ranching families, Betsy believes RFP gave them more resolution when trying innovative things. They connect to an international community through the RFP 'Executive Link', which includes their Australian partner organisation, "Grazing for Profit." The Executive Link meets three times a year, disclosing details across the entirety of their operation to their small peer groups, for all to delve through. After these meetings, ranchers leave inspired with new action plans that cover the full gambit of ranching life from enterprise choices, finances, ranch structures, stock classes, family dynamics, succession and marketing. As a group, they hold each other accountable. It's an extraordinary tool for success and to break the habits of a lifetime. For the Indrelands, it meant their trials and experiments didn't create a sense of loneliness and isolation, instead it has deepened their connection and sense of contribution to a global pool of knowledge.

They keep excellent records and have been tracking their soil and pasture changes, monitoring brix, soil minerals, biology and plant tissue tests to ensure they're heading in the right direction. Their irrigated meadows are bacterially dominated, with 'sleepy' soils on the range, a fairly typical finding in the Mid-West rangeland. They have soils with 100% Base Saturation, high Ca and Mg. Early testing revealed trace element mobility issues (Mn and B), low sodium and low nitrogen. Their irrigated meadows contain a diverse mix of introduced grasses and alfalfa, with yields that had been struggling due to tight soils and poor infiltration. Field observations and leaf tissue tests showed the hay was of average quality, providing lots of feed for insect pests, like alfalfa weevil and alfalfa flea. Taking weed tissue tests and visual soil assessments provided keys when identifying those limiting or enabling factors, low fungal activity and low boron and manganese.

Based on the testing, a bio-stimulant blend was prescribed which included trace elements, fish hydrolysate and humic acid. Before we began a regenerative programme, the alfalfa fields were average performers. Yields had begun to slowly decline. This table shows the measured changes in the alfalfa in a wet chemistry plant tissue test. The 'supreme' column is the base levels we are aiming for to sell and market this alfalfa as the highest quality. The conventional alfalfa was receiving standard soluble NPK fertilisers.

DM Basis	'Supreme'	Bio Alfalfa	Control Alfalfa
Crude Protein	>22%	29.7	**21.9**
NDF	<34	28.5	**37.5**
ADF	<27	26.7	**33.9**
TDN	>62	70.1	**62.4**
RFQ	>180	222	**155**

Chart showing the difference in forage tests between optimal 'supreme' tests for alfalfa, Indreland biological treatment and the control receiving NPK fertiliser.

These initial results were startling; this one-off treatment had effectively lifted feed quality by 43%! Livestock producers can see this result directly by observing animals; they will spend less time eating and more time lying down (unless they are a horse with no off button!).

The improved quality will mess with your feed budgets, if you're in a feedlot. The cattle effectively are now getting the same amount of nutrition from two mouthfuls, as they once got from three. This improvement in nutrition is retained in stockpiled Winter feed, increasing the valuable protein and energy levels. Brix levels doubled from 10 to 20°, all for only a $20/acre investment. These fields also increased 1 ton in yield lifting from 2 ton to 3. Working out the numbers on this, the additional ton yield pays for a further 5-10 acres of application for the following season. Quality hay like this, will store better and when markets become educated, will command extra at sale point. One hay producer in New Zealand, producing a high brix, solid stemmed Lucerne (alfalfa) for racehorses; found that once horses got a taste of his wares, they wouldn't eat any other hay. He could then dictate the market prices. Excellent.

In 2015, Roger repeatedly struggled to use a conventional sprayer, which blocked with any coarse materials. In 2017, he set about designing and building a slurry sprayer, based on the advice from brilliant New Zealander and Utah native, Steve Erickson, at Chaos Springs. The sprayer cost less than $5,000 USD, including labour, to build. It can pump huge volumes (300 gallons or 1200 litres/min) directly out of irrigation ditches or troughs.

The first time Roger used it, it paid for itself and more. One-day while cutting for hay, a neighbour drove past gesticulating wildly to a spot behind him. Flames were just beginning to lick across the field, started in the combination of hot rotary disc cutter and paper-dry grass. Fortunately, the sprayer was nearby. Roger quickly filled the tank from an irrigation ditch and within minutes had contained a 10-acre grass fire, one that could've cost the Indrelands their annual hay crop and Winter stockpiled feed at a cost far more than their original sprayer investment.

A slurry sprayer, is a coarse-nozzled spray unit, driven by an open diaphragm or trash pump. These sprayers are designed to reticulate liquids and keep solids in suspension. One advantage of the unit is that there are few points for blockages, due to the simple interlocking systems and no complex mixers or compression points. I prefer round tanks, rather than oval or square tanks. These round tanks can create vortexes, which adds another dimension of benefit and reduce sloshing --an essential element on uneven country. A single 5 mm nozzle (1/5 inch) can cover over 16 meters (52 feet), so cropping operations can put 2 nozzles onto a boom and cover 32 meters. They are great for spraying sieved compost, live biology and seed, with large droplet sizes at surprisingly low pressures.

With adaptive grazing practices and Winter calving, just a dark memory, there is now ample stockpiled forage for Winter. Hay cutting had stopped being a priority until 2017, when hay prices were just too delicious to resist!

The Indreland Ranch includes 400 acres of low-lying rolling country. These areas with their deeper topsoil, were traditionally used for oats and barley. What little organic matter remained, after historic grazing cycles, was rapidly oxidised or blown away. As a result, the farm ground degraded into lifeless, structureless clays and silts, dominated by early succession Yellow alyssum (*Alyssum alyssoides*) and non-mycorrhizal-species, Field pennycress (*Thlaspi arvense*) and lupines. It was an area Roger and Betsy initially left from the soils programme, since it was so degraded. This is a good strategy. By focusing on better-performing areas, or areas that do get moisture, subsequent lifts in quality and performance can help fund the cost of improving poorer areas. Through lifting the carrying capacity in their irrigated lands, they were able to lift livestock by 25%, numbers they continue to build upon.

In Autumn, 2017, the farm ground received its first inputs in 30 years. Using a direct drill, Roger mixed 2 pounds of dry vermicast and 2 pounds of Kelp meal with 12 pounds vetch and 50 pounds of Winter rye. As an additional experiment, he also added sunflowers, because he could and that's what experiments are all about, even when your "know-it-all" coach thinks you're crazy. The 2017 Summer had been a scorcher; over 300,000 Montana acres, had been on fire. Fortunately, Fall did bring some germinating rains. The sunflowers grew to 3 inches, before they were knocked out by frosts. It may not seem like much, but even 3" of growth and a wee taproot, would have some benefit to a land that had been growing nothing. Any cover is better than being caught out with a bare bottom.

Come Spring melt and warm sunny days, the vetch and rye sprung to life. Excited daily updates were received on their growth status. The 2018/19 seasons have been one of the best South-Central Montanans can remember, with some good regular rainfall keeping the covers growing. Even by late Summer, the base out there stayed green. Quite frankly, with the mud, it feels more like New Zealand than the yellow dusty country I have come to love. It's been a fantastic year to start on new ground. The cover crop grew to over a meter high in places and only 40 of the 400 acres were grazed. The rest has been left to self-seed and give the soil some well-needed rest and recovery. This land was Winter grazed and the seed that has now dropped was trampled to germinate in 2019. Roger has never seen this kind of growth on these lands and has been incredibly encouraged to expand their seeding/vermicast acreage. During the growing season, 100 acres received a biological boost. Leaf and soil tests showed major imbalances, a mix of minerals and biologicals were sprayed. A late rain showed the line where the foliar treatment had gone in, as a new flush of growth happened under the treated areas. Leaf tests showed that the treated plants were responding positively to nutrients, setting them up for better quality seed set, forage and carbon drawdown. After years of being treated like dirt, these lands are now flourishing.

One morning while Kate and I were moving cattle, we saw Roger from across a large pasture double over. He didn't stand up again and Kate began to worry. Was he having heart trouble? We called his phone. No answer. And no movement across the field. Suddenly, we saw him leap onto his ATV, arms flying as he zipped over to us. He had a broad grin and was full of contagious excitement. Beetles! He'd been trying to film a dung beetle he'd spotted rolling a ball of manure, something he'd never seen on the ranch before. The minute he'd stopped to watch; the dung beetle played dead. Roger remained motionless in the hope the beetle would set to work again, ignoring any phone calls in the process. He kept his camera primed, until the dung beetle slowly unfolded its legs and promptly flew off!

Since the management changes have been implemented and plant nutrient levels have been lifting, diversity has naturally returned. A friend of Kate, a young grassland ecologist in the making, has been out flipping cow pats. At last count, there were 6 different dung beetle species, including the escapee roller. With the taller pastures and Winter cover, more biodiversity is flowing in. Sage grouse (*Centrocercus* urophasianus) and Hungarian partridges (*Perdix* perdix) are now common sights, as they bob across the lawn with a trail of fluffy young. Both species look very tasty and with the voles and moles, the fox population is booming. One mum and dad fox set up camp on a knoll opposite Roger and Betsy's bedroom window. This Spring, we all watched in wonder as not 4 or 5 cubs, but 7 fuzzy balls tumbled out from their den. They provided entertainment for weeks. It's always a pleasure to be around families who rejoice in the diversity of life and see the benefits predators bring to the land.

The principle: "Without measurement there is no management" is demonstrated by all the Regenerators I speak with. This measurement may include observations, which Roger and Betsy both excel in. They have a comprehensive soil testing kit, which includes materials to photograph and monitor transects.

Through monitoring changes over time, Roger says, "We have gotten into the habit of tracking those things and observing results from a different perspective from which we had looked at them before." The actions I've suggested with the family have "had a positive result, whether it is an increase in Brix, dung beetles or earthworms." Infiltration improvements have been dramatic, with a 3-to-4-fold improvement in just a few years. When you ask Roger how much rainfall he got, you'll get a sassy, "Why, all of it, of course!" At least that's the goal. They were put to the test with a 6-inch storm event (2 inches were hail) in 2 hours, flash flooding causeways and the road to the north of the ranch. The next day, they had no issues driving across the ranch to survey the damage. Soil structure for the Indrelands has been one of the more notable differences. This structure differs across the fence. Driving a pick-up or side-by-side is easy going...until they open gates to drive across neighbouring properties, often to haul out bogged down neighbours!

The Indrelands are part of a unique soil carbon scheme, the Montana Grasslands Carbon Initiative. Driven by Western Sustainability Exchange and Native Energy, a carbon credit provider, the programme pays ranchers up front, for practice changes known to improve soil carbon, such as adaptive grazing management, range riding and avoiding tillage. These ranchers are using the funds to improve water systems and fencing. They submit grazing plans to the project and attend workshops on methods to improve soil health. The initiative has received a lot of positive attention from the community and the voluntary market. I'm not a fan of carbon offset markets personally, as the benefits from building soil carbon rewards producers directly. However, hiring the producers, who are responsible for most of the land, to improve ecosystem services, is something I am happy to get behind. As the U.S. soil guru Abe Collins says, "we are building the largest infrastructure project in the world." To achieve such lofty goals, the people on the ground need to be hired to build the system. Unfortunately, building soil health doesn't catch the eye, like a CO_2 scrubber, foodbank, dam or a bridge. The effect, however, is far more profound and effective than the 'ambulance at the bottom of the cliff' approach.

Many producers tell me they don't have time for monitoring; however, it's the monitoring which is going to give you *more time*. Try to create a simple system that becomes a habit. The most successful producers I know, carry a small notebook, a refractometer, garlic crush and a temperature probe by their truck's gearstick. When driving through a gate, drive an extra 10m (32'), put the moisture meter in the ground (away from the track) and as you walk to shut the gate, grab a few handfuls of grass, put the sample into your garlic crusher, look at the sample, shut the gate, walk back to the truck, record the temperature and drive off. This would be the same for horticulturalists. When you're walking down the rows, take around 20 leaves from different plants. Include pH sap readings in this sampling too. This will add all of a minute to your routine and a wealth of information as you build a picture of your place.

You will be sampling at different times of the day in different climatic conditions and different growth points. All this information is going to help you manage more decisively and build confidence that you're heading in the right direction. If you're *not* heading in the right direction, this information will point you in another direction. Plants are stressed? Take actions to support health, without losing production. When Brix is highest, cut for hay. For dairy and lamb fattening, if Brix is low, miss this field in the rotation. Low and sharp? Consider potential nitrates, don't graze and, if possible, apply a spray with humic or milk products to mop up the nitrates before weeds germinate. If your plant pH is low, try applying an alkaline spray like milk, liquid calcium, or seawater. Once crop and pasture health improve, you'll have more time on your hands anyway, which I'm sure you'll find ways to fill!

It is a constant amazement to me, how most producers don't take enough photos (although I'm just as guilty of this.) In 2000, I moved to my father's farm in the beautiful Bay of Plenty. The Bay of Plenty is as good as it sounds: deep, rich, young volcanic soils surrounded by sparkling seas, sheltered from cold sub-Antarctic winds, by bush-covered mountains. It's an area south of Auckland, NZ, famous for huge avocados and kiwifruit. At the time, Dad's farm was infested with the noxious weed Ragwort (*Senecio jacobaea*). It grew so high and dense, that cattle would disappear when they lay down. Ragwort is also known under more descriptive titles, such as Stinking tansy and Mare's fart (seriously?). We used to hand pull ragwort by the trailer load. Unfortunately, it contains alkaloids, which make me feel pretty ill (my name for Ragwort isn't suitable for print.) When locals asked where we lived, we'd reply, "On the yellow farm up Wrights Road," and they would knowingly nod. With bio-controls and the soil programme, Ragwort is a thing of the past. The only evidence I have is a single photo taken of the family out for a lovely picnic surrounded by a 5-foot-high wall of yellow weeds.

WHY MONITOR?

Benchmark: Know where you are starting from: Are your management objectives taking you forward or backwards? Identify early warnings.

Manage: Take action in response to indicators; Guide management of livestock and apply nutritional sprays or bio-controls.

Evaluate: When management strategy changes are needed to better meet identified objectives.

Record: Provide a record of environmental and resource conditions, events and management practices.

Inform: Provide information to inform management; grazing choices; species suitability; water management; Are bio-stimulants or fertilisers providing bang for their buck?

Warn: Early warning for practices which are declining soil health.

Track: Track changes over time; provide a record that can help to secure leases, partnerships or investment.

Proof: As your 'smug' test to put your money where your mouth is!

Long-term monitoring for soil carbon is typically repeated on a 4 to 5-year basis. More dynamic measures like Brix and temperature can be carried out quickly through the season.

See Appendix for what other soil and plant benchmarking tests you can take.

11

Read Your Weeds

"When you look at a field of dandelions you can either see a hundred weeds or a thousand wishes." Unknown

It's time to get excited about those weeds! For those of us interested in listening; they are here to tell us a story. A story that may be multi-generational, environmental, or even a reflection of your management. I rejoice in people's weeds, as they offer clues and insights into soil health and management and this all equates to opportunity!

Before we dig into this topic, I need to define what I mean by a weed: valuable indicators and storytellers. A weed is commonly defined by the highly subjective "a plant growing in the wrong place," which leaves a lot of room for what is, or is not, a weed. Consider different attitudes towards Lamb's quarters or Fat-hen (*Chenopodium album*). It can provide high value animal feed and is commonly grown in Africa and Asia for seed consumption and as a spinach substitute. In cultivated crops, however, Lamb's quarters is competitive and can lead to significant and costly yield decreases. Is ryegrass a weed? Dairy farmers value it for milk production; however, in broadacre cropping, ryegrass creates a persistent headache. In some environment's roses, eucalyptus and pine trees are considered weeds.

Over thousands of years, we have harnessed this competitive weedy spirit, to domesticate many of our staple crops, like wheat and rye. Many of these crops are annual species that flourish in disturbed ground--an adaptation they share with many "weeds."

It's not uncommon to hear people complain about their neighbours allowing weeds to go to seed, indignant at the invasion across their fences. The weed seedbank in the soil is massive; an individual Lamb's quarters plant, can produce over 600,000 seeds in a season, seeds which may remain viable in the soil for over 40 years. Seeds from some species can be short lived, such as Russian thistles, which only remain in soil for a year, while other plants are made of much tougher stuff. An on-going study in Michigan involving 20 common weed species, showed that after 120 years lying dormant, seed from Moth mullein (*Verbascum* blattaria) was still germinating. As with drugs and cancer, we will never win the war on weeds by being reactive in our thinking. We can, however, go to the root cause and change the signal that causes a seed to germinate.

Regenerators report that with changes in grazing and/or soil management practices, native species considered rare or locally extinct, reappear. Farm researchers in New Zealand measured around 1,200 clover seeds in a square meter of soil, yet not a single visible clover growing above ground. There is a substantial seed bank waiting for a signal for germination.

Spending time in the field to observe weed patterns, is invaluable in our diagnostics; what signal are you sending your plants and how can you change that signal? The health and diversity of plant communities speaks volumes about underlying soil conditions and management. The first steps are learning to listen and read the basic drivers for weed germination.

There are six main (and related) reasons why weeds germinate:

1. To colonise bare soil.
2. In response to low organic matter.
3. To open up compacted soils.
4. In response to mineral availability.
5. Microbial stimulation.
6. As a safety valve for toxins.[89]

1. Bare soil colonisers

Nature detests a vacuum. The conditions which create bare soil may be due to natural events like fire, volcanoes, erosion, or human disturbances such as cultivation, overgrazing and land clearing. These bare exposed soils are vulnerable to being eroded, sunbaked and sucked dry. In response, nature sends in her defenders to protect against these injuries.

Often these plant species have a scrambling or prostrate growth form to quickly cover the ground. Examples of plants that fall into this category, include Scrambling fumitory[9] , Purslane, Spotted spurge , Caltrop Field bindweed and many more! Some of these early colonisers have shallow rooting systems and the ability to produce an overwhelming number of seeds, such as the invasive Cheat grass. Countless species have adapted to colonize disturbed areas and assist soil building in a process coined 'plant succession'.

The early colonisers, include moss and lichen on rocks, as well as the complex communities called Cryptogams (*Cryptogamae*), (see box). Many graziers consider Cryptogams a weed, however, in the absence of a living plant root, they are essential soil protectors.

Management to meet graziers' goals, must focus on reducing bare soil and increasing topsoil development; over time these basic Cryptogam communities shift away from low quality annual weed grasses, towards denser higher quality plant communities.

In horticulture and vineyards, herbicides are used to maintain bare soil under plants. This practice is the opposite action for optimal vine and tree health. Some bare soil colonisers are quickly becoming herbicide resistant, as conditions favour weeds. Mowing or introducing livestock, is a preferable understory management tool. Commonly, what is seen as competition stress from weeds, is due to other factors, such as allelopathy or weeds feeding primitive biological communities.

---

[9] Many weeds have the same common names around the world, to ensure we're on the same page latin names are included as a footnote on each page. Scrambling fumitory (*Fumaria spp*), Purslane (*Portulaca oleracea*), Spotted spurge (*Euphorbia maculate*), Caltrop (*Tribulus terrestris*), Field bindweed (*Convolvulus* arvensis) Cheat grass (*Bromus tectorum*).

Cryptogams are biological communities that include lichen, cynolichen, algae, fungi, bryophytes (liverworts and mosses) and cynobacteria. They are vital colonizers in environments that have little topsoil. Through cunning methods, they protect themselves from harsh elements and when conditions are right, they use powerful acids to etch minerals for growth. They are one of the most understudied (and undervalued) plant communities. Cryptogams are found across all environments, wherever plant cover is low, including hot deserts and icy regions, like Antarctica. They have an intimate relationship with soil surface and help to create the crusts that form the fabric for higher plant communities to establish. These crusts capture dust and protect against losses of soil, carbon, water and UV rays. They sit at the base of the trophic pyramid, providing food for soil insects and larger herbivores. In dry environments, they remain dormant until dew or rainfall stimulates them to photosynthesise. Cryptogams are significant sinks for global carbon, with estimates ranging from 2.1-7.4 billion tonnes/year. This equates to around 7% of the carbon cycle from land plants! In nutrient limited environments, like deserts, cryptogams are significant sources of nitrogen, one study revealed they could be contributing as much as 100kg/ha/yr (100lbs/ac), nearly half of the global nitrogen cycle! Depending on the structure of these Cryptogram communities, they can either assist water infiltration (when rough), or if the communities have smooth surfaces, they can stop or slow infiltration.

Cultivation to create bare soil is a destructive practice, breaking up the vital soil aggregation and microbial communities, particularly if not done with biological considerations. The negative impacts from cultivation were heavily publicised in the early 20th Century by ecological pioneers, such as Edward Faulkner, Louis Bromfield, Newman Turner and Lady Eve Balfour.

They were advocates for the incorporation of cover crops and judicious use of shallow cultivation. After decades of building upon these approaches, there have been advancements and understanding around roller-crimping cover crops, the potential for no-till organic crops is becoming an increasing possibility, in some climates. Using methods to manage the other weed germination signals, will also help reduce plant competition.

Actions for bare soil

Avoid at all costs!

Address management issues that create bare soil.

Mow or crimp, rather than spraying or cultivate.

If you can't avoid, then mitigate harm by dripping humic acids when cultivating, adding organic matter and quickly getting seed back into the ground. Shallow incorporate cover crops. Add a carbon-based seed treatment to give new plants a jump start: seaweed, humic acid, compost/vermicast extracts. MF if needed.

2: Low organic matter (OM) plants

There are a vast number of species in this category, which also include many of the bare soil species. It is possible however to still have ground cover on low OM soils. The species prevailing in low OM soil, often have adventitious or deep-penetrating roots; think of species you'll find growing in the cracks of sidewalks or in gravel on the sides of roads, like Fleabane[10], Dandelions, Cape daisy, Knapweed, Mayweed , Hawkweed and more. These species use a variety of methods to increase resource availability in low OM environments, effectively priming the soil to release bound nutrients. They kick-start the carbon and soil building processes, capturing dust around their base and upon their death, they contribute to the OM pool.

The scourges of the American West--Leafy spurge, Cheatgrass and Spotted knapweed--all alter their soil communities to their own benefit and prevent

[10] Fleabane (*Erigeron bonariensis*), Dandelions (*Taraxacum*), Cape daisy (*Osteospermum*), Knapweed (*Centaurea)*, Mayweed (*Matricaria*), Hawkweed Leafy spurge (*Euphorbia*), Spotted knapweed (*Centaurea maculosa*).

succession to native grasslands. Both Knapweed and Spurge, offer beneficial services, as excellent mycorrhizal hosts, inoculating soil with spores. They have higher respiration rates than the native grasses, using this advantage to increase the turnover of carbon to their roots and soil. Through their root exudates, they encourage a proliferation of carbon-loving bacteria (copiotrophs), archaea and soil diseases, to outcompete their neighbours.

The plant process of rebuilding OM is not *quick*; it can take many decades to lift soil carbon to levels more favourable to the higher advanced grass species. As land collaborators, our focus is to ensure that we are always using practices to build, not degrade soil organic matter and do these outlaws out of a job!

<u>Actions for low OM soil</u>

Spread, feed and add any organic materials you can find, from rotten hay, straw, composts and biochar to mulching materials. Manage for maximum groundcover, animal trampling, herbal leys, cover crops and ultimately increase your plant Brix levels!

3: To open up compacted soils

Soil compaction may be due to mineral and/or microbial imbalances or management practices such as; overgrazing, tillage, soluble nitrogen etc. Many weeds provide an aboveground 'tell' that soils have become compacted. When surfaces become compacted or crusted, this reduces a soil's ability to breathe, providing the germination signal for species such as Clubmoss[11], Dock, Buttercup, Thistles, Pennyroyal Rushes and Sedges. In low oxygen or waterlogged soils, many plants will die, but not the Rushes, Sedges or Reeds. They have adapted to live in anaerobic soils, by pumping oxygen down through their roots to transform nutrient availability. If you see these species growing on hillsides, suspect that soils are not breathing, rather than waterlogged.

---

[11] Clubmoss (*Lycopodiopsida*), Dock (*Rumex)*, Buttercup (*Ranunculaceae*), Thistles (*Scolymus hispanicus L*), Pennyroyal (*Mentha pulegium*), Rushes (*Juncaceae*) and Sedges (*Cyperaceae*).

Dig a hole and look at weed root systems: do they stop or change direction through compaction layers? These soil breakers are plants which have amazing rhizomes or tap root structures, which open compacted soils.

Many are indicating that calcium is not functional. Deep tap rooted broadleaf weeds, such as Dock[12] and Canadian (also known as Californian) thistles (which may have root depths up to 20 feet deep) are providing a service by opening-up tight soils. They transport nutrients from the subsoil and create channels for air and water. Interestingly, Canadian thistles have specific anaerobic bacteria that live on their roots. When soils are healthy and well aerated, these conditions will not suit their bacterial partners.

Actions for Compacted Soils

Plant multi-species cover crops with deep tap roots, such as Safflower, Sunflower, Tillage radish, Lana, Woolly pod vetch, Yellow sweet clover, Berseem clover, Cowpea, Sorghum (*Sorghum*) etc.

Mechanical interventions: If you want to speed up opening compacted soils, there are good aerating or ripping tools on the market. If you know functional calcium is low, a litre/ha (1 pint/ac) of liquid lime or liquid gypsum dripped down the rips will help with aeration. Small amounts of humic acid, molasses or sugar, will help feed the microbes and encourage roots down the rip lines.

Always rip and drip with a carbon!

Then you need to get to work on addressing why you have compacted soils! Is it minerals, microbes, OM, management or mindset?

---

[12] Safflower (*Carthamus tinctorius*), Sunflower (*Helianthus*), Tillage radish (*Raphanus sativus)*, Lana (*Pristiophorus lanae*), Woolly pod vetch (*Vicia villosa ssp*), Yellow sweet clover (*Melilotus officinalis*), Berseem clover (*Trifolium alexandrinum*), Cowpea (*Vigna unguiculata*), thistles (*Cirsium arvense*.

4: In response to mineral availability

When mineral availability is poor, specific plants have a competitive advantage over their neighbours. Through a process called the 'rhizospheric priming effect' (RPE), plants can influence mineral and carbon cycles[90]. They often have deep roots or different root morphology enabling the species to mine minerals; we call these plants the 'dynamic accumulators.' Through exudates and signals to microbes, they can kick-start the soil building processes and lift functional minerals. The previous studies mentioning Spurge and Knapweed, also measured changes in soil mineral profiles, showing increases in pH, potassium, ammonium, nitrates and phosphate. This mineral release process helps to increase their competitiveness against many native grasses. There are some weeds that love nutrients in excess and have adapted to grow in environments with toxic levels of sodium, cadmium, lead, potassium and zinc.

High available potassium and low phosphorus (as determined in a Reams/Morgan soil test), provides the trigger for broadleaf weeds. These weeds include Dandelions[13], Common plantain, Black nightshade and Inkweed.

A commercial composter once phoned for help. His compost was infested with weeds, specifically Black Nightshade and Inkweed. I suggested they were responding to an imbalance between P: K, recommending that he add guano phosphate and turn the compost. What then germinated on his compost, were the soft grasses and clovers. He had changed the germination signal from broadleaf weeds, to species he wanted to grow.

Generally, primitive weedy grasses indicate low functional calcium (and bacterial dominated soils.) This term 'functional' is important, as even on limestone soils, you can have low available calcium. This functionality is driven by water and active fungi. It's perhaps of little surprise, to see that so many weed species indicate a breakdown with fungal life.

One way you can reveal if your weeds are dynamic accumulators, is with a plant tissue test. Sample when your grasses and weeds are at least 3 inches high and before they flower. Sample at least 20 plants. If you're sampling thistles, use a pair of heavy-duty gloves! Also, sample a grass/crop alongside your weed as a comparison.

---

[13] Dandelions (*Taraxacum*), Common plantain (*Plantago major*), Black nightshade (*Solanum nigrum*) and Inkweed (*Phytolacca octandra*).

The table on the following page, shows a leaf tissue test taken on Cape weed[14], a prostrate broadleaf in the Sunflower family, on a Western Australian farm. It's pretty, until you see it choking out grasses at a rapid pace in response to poor soil management. Originally from South Africa, you'll find it densely covering areas in New Zealand, Australia, Italy and along coastlines of the United States. In Western Australia, it may be the only thing you'll see growing in Summer and grazed early enough, it can provide high 'protein' feed.

Note the underlined nutrients in the table: these are where the weed notably differs in nutrient values, compared to the plant we would prefer to grow. Of interest, the weed tissue test was supported by soil tests which showed low Ca, Na, Zn, B and HIGH nitrates. These dynamic accumulator species will prime and release minerals which are low in the soil, the only exception to this pattern is high nitrates and excess sodium. These results helped to inform our decision on what nutrients to apply. The recommendation included a fine lime (calcium), guano (P), boron, zinc, cobalt, humate, sugar and sea minerals.

Actions: Test your weeds against your preferred crop and soil tests. Is it a functional deficiency (bank card) or a total deficiency (bank account)?

Address major mineral imbalances, if feasible.

Use small amounts of mineral inputs (with carbon) to provide a catalyst to shift mineral availability.

Nutrient		Units	Ryegrass	Capeweed
Nitrogen	N	%	2.57	2.18

[14] Cape weed (*Arctotheca calendula*),

Phosphorus	P	%	0.21	0.24
Potassium	K	%	2.39	2.30
Sulfur	S	%	0.18	0.18
Calcium	Ca	%	0.46	**1.43**
Magnesium	Mg	%	0.24	0.32
Sodium	Na	%	0.16	**1.17**
Copper	Cu	mg/kg	6	9
Zinc	Zn	mg/kg	16	27
Manganese	Mn	mg/kg	47	59
Iron	Fe	mg/kg	60	88
Boron	B	mg/kg	4	**39**
Molybdenum	Mo	mg/kg	0.5	0.4
Cobalt	Co	mg/kg	<0.1	<0.1
Crude Protein	Ratio	%	16.1	13.6
Nitrate	N	mg/kg	62.6	**133**
Ammonium	N	mg/kg	686	407

Table: Plant tissue tests comparing weed and preferred grass – rye.

5. Microbial stimulation

Most plants are telling you something about microbiology.[91 92 93] Ever wonder why trees struggle to grow when you plant them into grass? Or native grasses struggle in mine restoration sites? Different plant groups prefer different microbial communities. Dr Elaine Ingham, at the Soil Foodweb Institute (SFI), has undertaken decades of work identifying what microbial community interacts with which kinds of plants. Her research holds true in the field[94]. The best way to look at what biological community suits your plants, is to consider what environment they would naturally grow in. Picture an apple tree growing in the wild, there is some dappled light and good leaf mulch. These environments are fungal dominated, you would see and smell mushrooms on the forest floor. SFI testing recommends a 1:10 ratio of bacteria to fungi in orchards.

In a healthy grassland ecosystem, the high value grass species, such as Tall prairie grasses[15], Rye, Wheat or Corn all prefer to grow in soils with equal biomass between 1:1 or 1:2 B:F. (Note that these ratios are based on the SFI method of testing, other labs such as those using PLFA will have different ratios.)

Nothing in nature is ever as simple or as linear as we love to imagine. There is a large element of random chance involved in who germinates and what the final community may be, based on moisture and climate. This disturbance timeline is only a general guide.

Consider that when soil is bare (or disturbed) bacteria are ever present; they're in the air and on every surface. The first species to arrive, can survive without soil--the lichen, mosses and cryptogams. Then the early colonisers move in, they have shallow rooting systems and scrambling growth patterns, to quickly gain a foothold. The deeper-rooted species open up cracks and release bound nutrients. As more dust, debris and organic matter accumulates, the scene is set for *Brassica* species and primitive grasses. As disturbance decreases, the advanced grasses move in and soil-building processes accelerate. As soils become less disturbed, they tend towards more fungal biomass, and in come the woody weeds, blackberry, roses and sagebrush. With even less disturbance, these soils become 'sleepy', providing the germination signal for shrubs. Over time, the trees begin to dominate. Landscapes may be affected by disturbances such as soil mineral availability, climate and altitude; a woodland, for instance, will never grow on the top of Mt Everest. Soil disturbances always shift soils towards a more bacterial state, while a lack of disturbance sends soils into a more fungal or 'sleepy' mode.

Early colonising plant species prefer and support bacterial-dominated environments. Often these annual species produce huge volumes of seeds, perpetuating their dominance and holding back succession. They can have lower animal feed values or be coarser than the more palatable 'advanced' grasses. Microbial testing under these grasses, such as Crabgrass[16], show high archaea and bacterial dominance[95]. Many successful invader species

---

[15] Tall prairie grasses (*Panicum virgatum*), Rye (*Lolium*), Wheat (*Triticum*), or Corn (*Zea mays*)

[16] Crabgrass (*Digitaria*), Medusahead rye (*Taeniatherum caput-medusae*) Ryegrass (*Lolium*),Forage radish (*Raphanus sativus*).

such as Medusahead rye and Cheat grass, reduce microbial communities and the N-cycle, making the area inhospitable for other species.

One strategy plants use to reduce competition is the release of chemicals, called allelopathy. Species such as Black Walnut trees and Ryegrass are well known to produce these suppressive chemicals. Producers can harness these plant chemical strategies to reduce herbicide use, by applying extracts that mimic those naturally produced. Some producers harness allelopathic properties by growing crops like Forage Radish for weed control in the following crop.[96]

When soil is disturbed through practices like tillage, certain weeds will germinate with a flash of light; however, studies of tilling at night, still found the same amount of germination. I believe this is a biological signal, as cultivation causes a spike in simple bacteria. Pugging (animal hoof impact), will also cause localized disturbance, creating bacterial dominated anaerobic zones. These feedback channels go both ways; plants signal to the microbes and the biology sets the scene for the plants. Can plants grow in less than ideal microbial communities? Yes, they can. However, they will be photosynthesising at lower rates (lower Brix) and more prone to stress, pests and diseases.

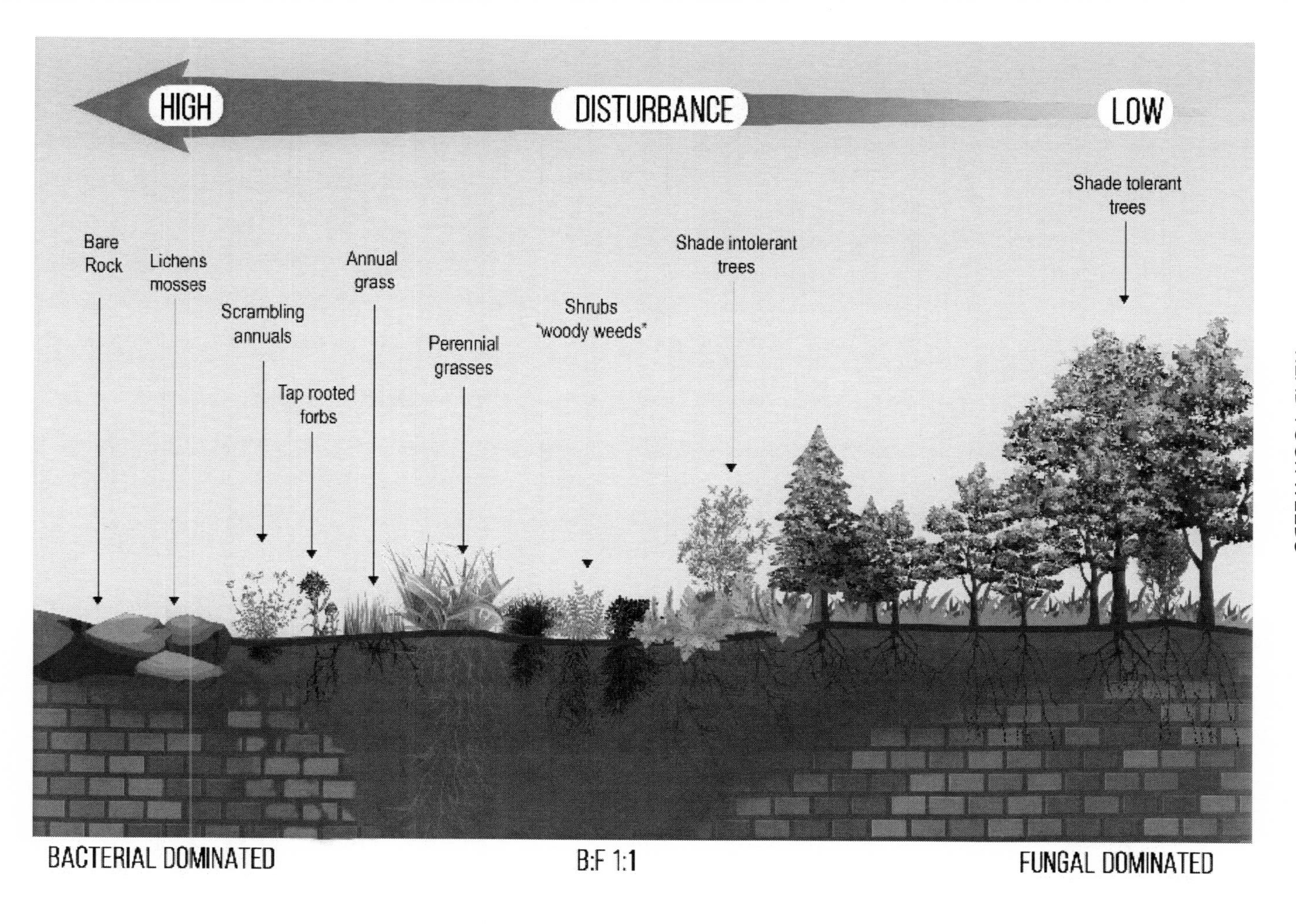
HIGH
DISTURBANCE
LOW
Bare Rock
Lichens mosses
Scrambling annuals
Tap rooted forbs
Annual grass
Perennial grasses
Shrubs "woody weeds"
Shade intolerant trees
Shade tolerant trees
BACTERIAL DOMINATED
B:F 1:1
FUNGAL DOMINATED

North of Yellowstone Park, ranches in the Tom Miner basin are home to the highest population of predators in the lower 48 states. There, grizzlies come down into the lower fields in search of Caraway[17] which provides a rich source of phosphorus, boron and energy, which bears use to fatten up for Winter. Caraway is a biennial plant in the carrot family Apiaceae, originally native to western Asia, Europe and North Africa. There are currently no known natural controls for Caraway in North America. The bears have quickly learned to dig up Caraway roots, in the process becoming accidental gardeners.

When bears dig up the soil, they create more bacterial conditions and bare soil, perfect for the germination of more Caraway seed. Some fields have over 30% Caraway cover. The ranchers don't want it; firstly, as it attracts bears into areas, which increases the chance of a dangerous encounter; and secondly, cattle don't like it so much. Our testing showed Caraway is a dynamic accumulator for phosphorus, boron and potassium (K). We measured 6% K in the plant tissue. When potassium is over 4% in forage and there's an imbalance of the other cations, there is an increased chance of bloat and metabolic issues in cattle. Cattle in their wisdom know this and will only pick a little to balance their diet. Comparing sites where Caraway hasn't gained a foothold in the basin, shows that these areas are fungal dominated with a F: B ratio of 4: 1. Caraway is a bacterial-loving plant. These less disturbed soils are just not providing the signal to germinate.

Brambles always seem to get people's minds all in a tangle around this idea of disturbance. Most livestock operators think woody-type weeds are early successional plants. Reflect where you commonly see these plants... it's where pasture sizes are large or stock density is low, or under fence lines, after forestry, or where land is abandoned.

There's a dynamic that occurs when cattle graze in large fields; they favour certain places and avoid the less appealing ones. This results in overgrazing in some areas and undergrazing in others. Dr. Richard Teague, Texas A&M, authored an excellent study showing that when pasture sizes are too large, only 39% of the land was covered by animals, leaving 61% untouched.[97] The

[17] Caraway (*Carum carvi*)

untouched areas begin to 'fall asleep.' In turn, this lack of biological activity triggers the germination and growth of these woody species. Many of these weeds are perennials, so shifting these dynamics back to grass takes longer to shift. To accelerate the shift, a disturbance is required such as fire, herbicide, mechanical invention and/or animal impact. Some producers speed up the disturbance and trampling in an area by feeding hay bales or placing a salt lick or mineral tub in the middle of a woody weed patch. Fire or mowing can shift the microbial community. However, this is only a temporary measure; you will still need to take actions to interrupt the germination signal. With livestock, putting in more fences and water and increasing stock density is key, otherwise these areas will revert again.

When fungal-dominated forestry or woody areas are disturbed, fungal numbers reduce, yet the soil will often be more fungal than ideal for grass or row crops. Take the example of establishing new grass after forestry; this may involve tillage, soluble fertilisers, herbicides and fire, yet weeds such as blackberry, rosehips, shrubs or poison oak still want to creep in. Consider species you see thriving in a woodland environment; some soft short grasses, Wild strawberries and flowering plants such as Hollyhock[18], Foxglove, Wormwood or Woolly mullein. When you see these woodland species cropping up in your fields, they're a good indicator that soils are becoming more fungal or "sleepy." These conditions often occur in dry areas, as fungi hold on better in the dry conditions than bacteria.

In New Zealand, the hill country invader, Hieracium, or Hawkweed, a low growing invasive flatweed, threatens production by outcompeting palatable grasses. These plants spread with inefficient grazing management. Thinking that the sheep were the problem, farmers removed grazing ...creating more weed issues. Removing any grazing, created sleepy soils, perfect for Hieracium, a fungal-dominated weed.

There are weeds that are either mycorrhizal (MF) hosts or non-mycorrhizal. Identifying if they're MF hosts or not, can provide you with powerful management insights. Over 90% of plant families have MF relationships, with varying degrees of dependence. Entire plant families have evolved to be non-mycorrhizal. There are always exceptions to the rules of course; some of

---

[18] Foxglove (*Digitalis*), Hollyhock (*Alcea rosea*), Wormwood (*Artemisia absinthium*), Mullein (*Verbascum spp*).

these families have one or two rebel species that will form mycorrhizal relationships.

Interestingly, in the most phosphate (P) limited areas, like the ancient landforms in Western Australia and South Africa, you will see the most incredible diversity of non-mycorrhizal species, like the ornamental Banksia, Leucadendrons and Protea. It is a 'bucket list' event, to drive north along the western coasts of Australia, to catch these species in bloom. These plants either develop cluster root structures or use other strategies to make P more available. These strategies can involve exuding high amounts of sugars or powerful acids out of their roots. Diffusion of phosphate in dry soils is poor, so these root structures can have an advantage. Root exudates from these non-MF species include phosphatase, phenolics and mucilage's. They also increase carbon by-products, such as carboxylates, which include complex heavy metals, aluminium and iron and in the process release bound P. This is how species like Brassicas[19] can acquire P and why Buckwheat and Lupins provide a great intercrop option for other crops. There are some species that do not have specialized root structures or chemical adaptations, such as Redroot pigweed and Nettles, but these species tend to grow in habitats with low competitive pressure and nutrient-rich zones. You will often see high numbers of these weeds in cultivated, fertilized or manured fields and commonly around stock camps or yards where there is a heavy loading of N, P, K. Many of these non-mycorrhizal species, have the ability to chemically attack mycorrhizal species and out-compete many of our commercial crops.

Typically, if you see a lot of these non-mycorrhizal weeds, something has impacted on the mycorrhizal population. Like the example we saw at Cottonwood Ranch in Nevada, dominated by Poverty grass [20] (a non-mycorrhizal species which was encroaching due to the water repellant, compacted conditions.) Tillage, overgrazing, waterlogging, some pesticides/herbicides and soluble phosphate fertilisers, also reduce MF. When non-MF plant species dominate, they're telling you they have the winning hand over the MF plants. Remediation options can focus on inoculating or feeding fungal populations in the soil. If inoculating, locally adapted mycorrhizal populations will be more beneficial than the limited number of commercially produced species.

---

[19] Brassicas (*Brassicacae*), Buckwheat (*Fagopyrum esculentum*), Lupins (*Lupinus*), Redroot pigweed (*Amaranthus*), Nettles (*Urticaceae*).
[20] Poverty Grass (*Juncus tenuis*),

Actions for microbially imbalanced weeds

Stop practices which kill your micro-herd!

Ask who am I gardening/farming/ranching for?

Take a microbial soil test.

Check Brix and see which plants are the happiest under your management.

Optimise grazing management. Avoid the use of spray strips under trees and vines- you are creating a bacterial zone under fungi-loving plants. Instead use under-row mowing or use thick mulch/fungal compost.

Use biostimulants, compost extracts, compost slurries, carbon inputs or management changes to push succession backwards or forwards to better suit your crops.

You can make liquid extract or inoculate your compost by sourcing litter, duff or soil from surrounding healthy grasslands or forested areas to economically reintroduce spores.

6. Safety valve for toxins.

There are several plants who act as a safety valve to remove high levels of toxins and nitrates out of the soil. They may also re-mineralise the soil, feed microbes and create groundcover, thus fulfilling all six roles at once. These plants are fast growing and often soft. They are nature's "repair plants" working to detoxify the soil.

Like everything in life, moderation is key. Around the world, there is increasing concern around nitrates in groundwater with causal links to cancers[98] and "blue baby syndrome." When babies drink formula milk made with nitrate-rich drinking water, the nitrates oxidise hemoglobin. This restricts oxygen in the blood, leading to comas and ultimately death. For livestock, nitrates lower cow fertility, cause abortions and can also lead to death. A visual test for nitrate poisoning is to look at their blood. Blood low in oxygen will be chocolate brown. Nitrate poisoning may occur from being fed poor-quality feed or from nitrates in drinking water, following frost events or a build-up of nitrates in pastures after days of cloudy weather.

In plants, high nitrates lead to an increase in pests and diseases. These nitrates can accumulate from producers applying rich nitrogen forms, such as fertilisers, chicken litter, poor composts, or from having biologically imbalanced soils, low in protozoa and nematodes.

A note on nitrates: most organisms, including plants and animals, struggle to process nitrates and nitrites. There are, however, a handful of weeds adapted to mop up excess soil nitrates. A signal for an excess of nitrates is the presence of the nitrate-accumulating weeds. They draw nitrates from the soil and into their plant structures. When they die, nitrate is returned to the soil in more complex nitrogen forms such as amino acids and proteins. The nitrate accumulators include Nettles, Fat hen, Foxtail Barley grass[21], Kochia, Nightshade, Cape weed and Russian and Milk thistles. Foxtail barley grass is interesting as it has a beneficial relationship with a unique mycorrhiza, which enables it to colonise high sodium and/or high nitrate soils.

You can confirm if the presence of these weeds is indicating high nitrates by using your refractometer. This provides an instant and powerful tool. Nitrate excesses are revealed with a reading of 3 ° or below and the line is very sharp. This reading indicates that proteins are not being complexed (free amino acids and nitrates) and sugar is not being sent to your soil microbes. As a result, pasture quality will be poor, and animals will scour (loose manure). Don't graze these pastures, you are losing production and worse-case scenario, you will also lose animals! In vegetable production, these plants will not store well and flavour will be poor or even bitter. Crops high in nitrates are more susceptible to pests and diseases.

Start testing the Brix level of your weeds and compare the results to your desirable plant species. You always want higher Brix levels in your crops than your weeds. If this is not the case, then potentially the current soil conditions, suit weeds better than the crop you are growing. This can be a real eye-opener; many people are farming/ranching for weeds!

---

[21] Foxtail Barley grass[21] (*Hordeum jubatum*), Kochia (*Kochia scoparia*), Milk thistle (*Silybum marianum*)...

A few years ago, while running a workshop at a sheep stud in New Zealand, the farmer hosting the event, shared that they were having difficulties with growing Chicory (*Cichorium intybus*) to fatten lambs. She found that the lambs struggled to gain weight, would scour and some would even die. In the field, the Chicory strike, growth and recovery was meager. When we measured the Chicory, it had a Brix of 3°, with a very sharp line--an indicator for nitrates. The fields contained very healthy-looking thistles and Black nightshade, a toxic weed. As no one was keen to sample the thistles for Brix, we sampled the nightshade. It had a Brix of 18° and the nightshade were shiny, healthy and happy. The management conditions were ideal for the growth of weeds.

The farmer had some raw milk available, so we trialed 3 different rates of milk, at 10, 20 and 40 litres/Ha (1, 2, 4 gallons/acre). Why milk? Well, it contains sugars, Ca, P, vitamins, minerals, proteins and lipids, and the minute you expose milk to the air--beneficial lacto-bacillus start to multiply! Forty minutes following the application, we re-measured the plants. The Brix level in the Chicory had increased to 8 ° and the line was now fuzzy: the plant was now photosynthesising at nearly 3 times the rate and feeding the soil microbes it needs. It also meant the farmer could bring in her lambs without risk. When we measured the nightshade... its Brix level had dropped to 6 °! With only 40 litres of milk we altered the plant/soil signals immediately. How long will this effect last? This depends on your base minerals. This operation had very low base saturation (BS) calcium, so the response only lasted for 3 weeks. In low rainfall environments with high BS calcium, this effect may last longer than a year.

In soils with heavy metal toxicity, or even radiation, there are plants and trees that can phytoremediate: pull the metals out of the soil to bind them in their roots, stems or leaves. These plant remediators include Willows, Silver birch, Flax, Hemp and Cotton [22],. These plants may return the metals back to the environment in a less toxic form, or be harvested for clothing, timber or biofuels. Mine sites have been investigating the use of plants that can draw metals out of the soil, to then harvest and extract the metals from the tissue.

This ability to accumulate metals needs to be considered when grazing animals in areas with toxicity, as *Amaranthus* and many brassica species, can

---

[22] Willows (*Salix alba L.*), Silver birch (*Betula pendula*), Flax (*Linum usitatissimum*), Hemp (*Cannabis sativa*), Cotton (*Gossypium*).

draw up lead, copper, and cadmium. The multi-talented Sunflower, has even been used to remove cesium and strontium from ponds after Chernobyl. They're also powerful for drawing out excess copper, manganese, zinc, arsenic and chromium.

I stumbled across another interesting dynamic, with the weed known as Milk thistle, or variegated thistle. The first time I encountered it, it was growing densely in a single gully on a new client's farm in New Zealand. When I asked the farmer what had happened there, he replied "it used to be an old dump site and we threw a load of car batteries in there." I've since noticed it growing in 'sacrifice' fields, where livestock are held following pour-ons or drenches, or behind barns where chemicals have spilled. After the 2010 earthquakes in Christchurch, Milk thistle started appearing on farms in the area. In a chance discussion with a physicist, I learned that when the earth's crust grinds together, it releases trapped radioactive gasses, specifically radon. I couldn't find much in New Zealand literature, but then I came across the same phenomenon in California along the San Andreas fault line--dense 1.8m (6') tall Milk thistle. This area is a hot spot for radon. What do we use Milk thistle for in human health? It's a liver detox. Your weeds are trying to communicate, and with observation and testing, we can learn to put them out of a job.

Actions: Use a refractometer to test who is happier--your crop or the weeds? Don't graze fields or harvest crops with a sharp Brix level below 3.

Nitrates can be mopped up with humates, milk, fish hydrolysates and vermicast.

Take a soil and plant test before buying land to ensure you don't have any nasty surprises lurking.

The Green Revolution came roaring in on promises of increased yields and the elimination of weeds, pests and diseases; for a few decades these promises yielded true. Today, modern farmers are at war, an arms race that is showing no signs of abating as more chemical interventions are introduced. Just as the haphazard use of antibiotics has created an epidemic of antibiotic resistant microbes, we see similar trends with other chemical controls. Chasing pure monocultures is the perfect recipe for a relentless march of herbicide resistant weeds across the planet.

The regenerators, Di and Ian Haggerty in Western Australia (WA), have observed a remarkable phenomenon with weeds. A block they recently purchased was seriously degraded from years of soluble fertiliser and heavy herbicide applications. Despite every chemical on the market being applied, herbicide resistant Radish, Melon[23] and Rye, were winning the war. When Di and Ian began to manage the land, sheep were brought in to graze. Crops were planted with worm extract and heavier rates of compost extracts were used (100 litres/Ha,10 gal/ac). Then they stepped back to let nature do the work.

In their 2nd year of transition, low successional grasses grew back in their Summer fallow period (in WA, crops grow in Winter). These grasses included Button, Kerosene and Windmill[24], typically considered as low palatable weeds in WA. To the Haggertys, however, these grasses meant valuable groundcover and a carbon pump for microbes over the hot, dry Summer months. In the 3rd year, something truly remarkable happened; higher successional native C4 grasses germinated across thousands of acres. This land had been terribly abused for over 60 years, yet the native seed bank was just waiting for the right signals to germinate.

---

[23] Melon *(Cucumis melo var.* cantalupensis).

[24] Button grass (*Gymnoschoenus spaerocephalus*), Kerosene (*Aristida contorta*), Windmill (*Chloris*) grasses.

Image: Prospect Farms during Summer 2018, native C4 Setaria grass growing in the furrows where vermiliquid was applied at seeding with a cash crop.

The grasses in the previous image look as though they've been sown down the furrows. The plants now growing are native C4 *Setaria* grasses, not seen in this area for well over 60 years, germinating right where vermicast/ compost extract had been applied. This is quorum sensing at work. In such low rainfall areas, there is a high risk that germination moisture may not arrive. With that in mind, the Haggertys choose not to risk spending money on cover crops. Seeing the results from actively altering the germination signal, has opened a door to the possibility; they can simply stimulate the seed bank which is naturally in their soil. What seeds sit dormant in most soils, never to receive a germination signal.

Ian still uses some herbicide selectively over his land, and what little he uses is made more effective by adding worm extracts. Through changing the biological signal, the herbicide resistant weeds are now becoming more susceptible. The worm extracts contain materials which, in my opinion, can break down the alkaloids and other secondary metabolites protecting the coating on the weed seeds, reducing herbicide resistance. This visual observation was backed up by sending in soil samples with seeds to do herbicide testing.

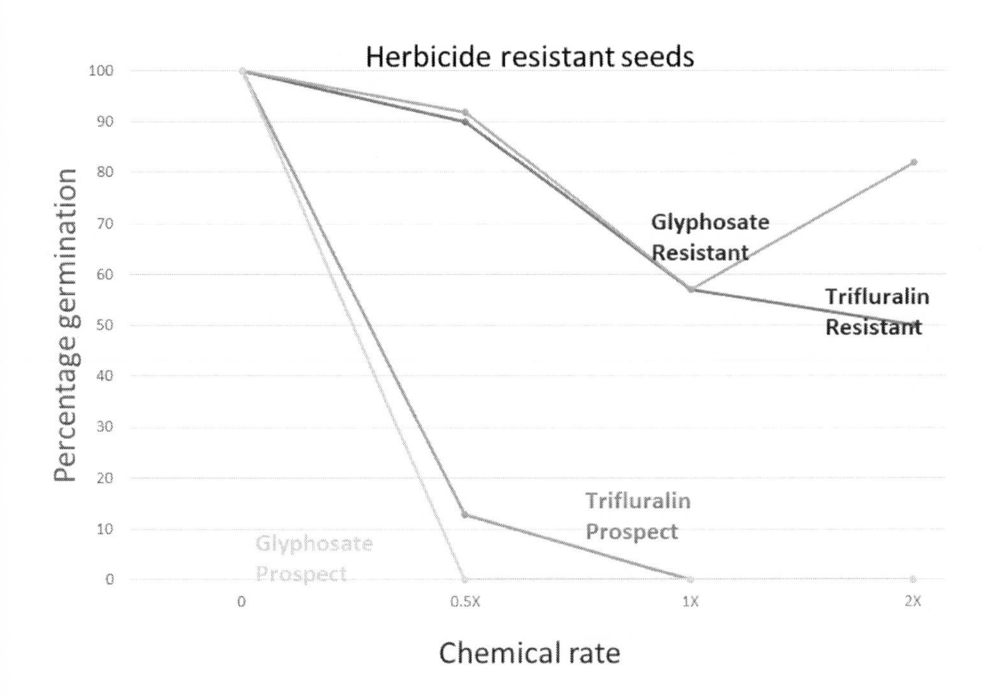

Graph showing the difference in herbicide resistance to glyphosate and trifluralin after 3 years. Prospect use the seed from biologically treated soils vs seed which had never been treated with their programme.

As the graph shows, in just 3 years, these persistent and troublesome weeds are now being killed at half the herbicide application rate. What is creating this dynamic has yet to be researched. What is clear, however, is that nature knows what to do, we just need to get out of the way!

Chemical weed sprays are effective in the short term at controlling weeds. Over time, however, these sprays create bacterial soil conditions that favour the growth of more weeds and tie up trace elements such as manganese and nickel; elements that plants need for defence. There is a multitude of mechanical and cultural management techniques for weeds.

For those of you who can't see how you can ever get away from spraying; you can buffer the side-effects of most herbicides through adding fulvic acid. Fulvic does an excellent job of stimulating the growth of beneficial microbes and it increases cell wall permeability. This means the chemicals (or foliar sprays) can enter the plant more efficiently, saving you money and supporting your soil health goals.

As the key drivers for weed germination are addressed and Brix level rises, you'll begin to notice that diseases and insects start to target the weeds. This is one dynamic I'm guaranteed to receive a phone call about, with a glee-filled producer exclaiming about insects moving from their crop and instead eating a troublesome weed.

Observing the whole picture is vital: dig holes, record changes, take tissue tests, note previous management or climatic conditions and set up some trials for yourself. This all helps to build your own knowledge bank. Understanding and taking action around the conditions your crops need for health, can save you on weed controls, crop losses and stress!

Action to cut herbicides: reduce **non-selective, non-residual** herbicides by 30% and add 1-part fulvic acid (or vermicast extract) to 4 parts herbicide (talk to the fulvic supplier to get absolute rates). Note: not all herbicides mix with fulvic, so do a jar test first, and test rates on a small area before rolling out. With so many different chemicals on the market now, fulvic acid may not work with them all.

How weeds grow, above and below ground, offers valuable clues to their soil-repairing role. Weed root systems can tell you volumes about soil structure and management. Integrate weed thinking with all the other soil health measures; management, microbes, minerals, OM and mindset. Mindset is the critical factor in the success with weeds. If you want to get rid of weeds overnight, it's easy... just stop calling them weeds! Now that you have more tools on how to read your weeds, it's time to take your eye off them and focus on what you want to grow instead!

Cultural weed control methods

- Cross graze other livestock species.
- Train livestock to eat weeds. There are great resources developed by Dr Fred Provenza on the Utah State University website www.behave.net.
- Inoculate seeds with compost/vermicast extracts or packaged biologicals to support crop health.
- Biocontrol agents: fungi, bacteria and insects that will target pest plants.
- Mow under trees or vines. Add more low-growing companion species.
- Address the key drivers for weed germination and lift soil health to set the scene for what you want to grow.

12

Good Bug, Bad Bug?

"Land, then, is not merely soil; it is a fountain of energy flowing through a circuit of soils, plants and animals." - Aldo Leopold, A Sand County Almanac

Taking a stroll through the 840 acres in New York's Central Park, urban dwellers and tourists gain a soothing respite from the turmoil of the city. Views down the main promenade are arched with American elms vividly recalling scenes from movies like *When Harry met Sally* and other blockbusters. The park contains over 20,000 trees, 280 bird species and a hum of insects. In 2003, this hum included the gnawing sounds of two invasive aliens: the Asian longhorned beetle and the emerald ash borer. These two pests have had a catastrophic impact, responsible for the deaths of millions of trees in North America, since their respective arrival in the 1930s and 2002. With their discovery in the park, New York City and 135 square miles surrounding the area, were put into immediate quarantine. The original infested trees, including their root materials, were removed and burnt. This was seen as the only effective control following infestation. With the iconic park's image at risk, the decision was made to aggressively defend the trees using the neonicotinoid, imidacloprid. If you sat under a tree in Central Park between 2005 and 2007, you were guaranteed a close personal encounter with this neonicotinoid (neonic), with over 14,000 applications used, during that 2-year spell[99]. Neonics, with their high solubility, were applied either as soil drenches or injected directly into the trees. They target specific insect receptors and so were considered far safer than many other insecticides on the market at the time.

In the early 1980s, there were three main broad-spectrum insecticides in use: organophosphates, carbamates and pyrethroids. With a heavy reliance on these insecticides, these chemical controls were becoming increasingly ineffective as pests developed resistance.[100] Pesticide resistance has been explained by adaptation in a process called hormoligosis, the theory being, that with sub-lethal exposure, insects adapt and evolve to resist the chemical. Many scientists thought these dynamics explained the increase in pest insect pressures observed *after* spraying. However, the story is far deeper and more complex than this. In vibrant and alive ecosystems, there are checks and balances in place that mitigate whole-scale vegetation losses. In healthy functional ecosystems, many of these so-called insect pests and disease organisms provide beneficial services. Just as weeds are here to tell you something, so too are pests and diseases.

During the 1980s panic, that growers would be unprotected against resistant insect hoards and with growing awareness around environmental and human health risks, new systemic chemicals entered the world. As these pesticides move systemically inside the plant, manufacturers argued they would only target chewing insects. By using substances marketed as being far less dangerous, growers could breathe a sigh of relief. These pesticides could be applied prophylactically through the growing season, instead of exposing people and bees through aerial sprays. These pesticides are now used indiscriminately around the world in seed treatments, mixed with irrigation water, injected directly in trees, or applied foliarly (hopefully after the bees have gone to bed).

The systemic pesticides include groups of chemicals such as neonicotinoids and phenylpyrazole (fipronil). Fipronil is commonly used around households to control fleas, termites and cockroaches under tradenames like Frontline, Goliath and Termidor. Neonics were developed in the 80s, by the dream team of German multinational pharmaceutical company Bayer and Shell. Chemically similar to nicotine, they disrupt the nervous system of insects, resulting in "mad bee disease" and death. As neonics travel throughout the plant, they expose the innocent bystanders, known as "non-target insects," through pollen, dew and nectar. A global analysis of 198 honey samples, found 75% of all samples contained at least 1 neonic.[101]

Pesticides are the most inefficient of all the agrichemicals. It is estimated that at best 1% of these chemicals reaches their target sites, as nearly all is lost to run-off, spray drift or degraded in sunlight.[102] In the case of neonics, only a tenth of the seed treatment is taken up by the plant, leaving the remaining 90% to impact on non-target species in soil, dust and waterways. Recent studies has shown that migratory birds ingesting, even low doses of neonics, become "anorexic," losing 6-25% of their body weight and have costly delays to their migratory patterns.[103] [104] A study released recently on waterway health in New Zealand, should be ringing alarm bells for all; between 2 to 6 different pesticides were found in 78% of streams sampled.[105] The organophosphate chlorpyrifos was found in most of these samples, this insecticide has a court order ban in the US and is banned for residential use in New Zealand and in many countries in the EU, including Germany. Ironically, this most widely used insecticide, is banned in its home country. Bayer continues to offer scientific assurances and shows dismay at the suggestions that neonics harm birds or bees as "Bayer cares about bees."

The timing of an explosion in neonic use in the mid-2000s, went hand-in-hand with the sudden collapse of bee, butterfly and bird populations. The Central Park treatments were celebrated as a success until, curiously, the treated trees began to turn yellow and lose their leaves. Closer inspections revealed a tiny spider mite, *Tetranychus schoenei*. Overnight, this mite, once considered a harmless herbivore, had turned into a raging beast, causing massive damage to the valuable trees. Initial assumptions were that the neonic wiped out the mite's predators, the lacewings, ladybirds and parasitic wasps, turning a shackled monster free.

This phenomenon is not limited to elms. Other researchers discovered that following neonic applications, mite populations boomed between 100-200%[106] in crops as diverse as corn, cotton and tomatoes. Mites are unaffected by the systemic pesticides as they lack the receptors that the neonics target. Measuring predator populations and other influences, did not explain why the mites began to produce nearly twice as many offspring. Researchers became curious. If it's not a lack of predation influencing the population growth, they wondered, what could be the cause? In a breakthrough study, they uncovered a cascade of changes to the genes inside the trees themselves. The activity of over 600 genes were altered with the application of a single neonicotinoid.[107] 600 genes! Many of these genes are responsible for cell wall structure, detoxification and switching on enzymes and phytohormones involved in defence. The neonics also increased the digestibility of nutrients, lifting available nutrition for the mites, resulting in an increase in the number of young. The insecticide created optimal conditions to weaken the plant and invite other pests to the table[108].

Many of the crude, broad-brush chemical controls have set agricultural systems up for the proliferation of pests and diseases. In a chemical arms race, it's the insect pests who are winning the war. For every 1 pest species, there may be as many as 1,700 non-pest insects[109] who have become the unintended casualties of this war. Insects provide a multitude of ecosystem benefits from pollination, nutrient cycling, decomposition and fueling the foodweb. The impacts from a looming 'insectaggedon,' the collapse of insect species, is broad ranging, far-reaching and potentially catastrophic. Although non-target species, like bee and butterfly populations are collapsing, the crop pest species are flourishing. There are now over 550 insect species resistant to pesticides, including insects that have evolved to consume the *Bacillus thuringiensis* (BT) toxin contained in engineered corn, cotton, soy and potatoes.[110] Despite an increasing complexity of chemical controls, pests still consume 18-20% of the global crop and are becoming increasingly resistant to the controls.[111]

With BT technology and targeted systemic chemicals, one could be forgiven for believing in the promised hype for a reduction in pesticides. Despite the benefits promoted by the seed producers, insecticide use has increased, not decreased, since BT technology was released.

In 2014, a public EPA memo stated, "published data indicate that most usage of neonicotinoid seed treatments does not protect soybean yield any better than doing no pest control."[112] Despite this information, these pesticides continue to be pushed upon producers around the world, as the gold standard in crop protection. Today half of soy and 79-100% of corn crops in the US are sown with a neonicotinoid pesticide.[113] A 2016 review, applying whole systems accounting to pesticide use, found the benefit ratio falls below 1. Which means that for every benefit pesticides offer, there are 99 costs. These calculations include environmental and human health costs. In the US alone, the direct and hidden costs of pesticides are estimated to be costing the US economy over $37 Billion USD every year.[114] A total rethink on pesticides is urgently required.

A 2018 study in the US corn belt comparing regenerative farms to conventional farms using insecticides, found ten times more insect pests in the conventional.[115] Yup, you read that right, where farmers were applying their full arsenal of insecticides, genetically engineered, plants and seed treatments, there were 10 times more insect pests.

That there is a relationship between chemicals and pest pressures is not new science. Sixty years ago, agronomist Francis Chaboussou, from the French National Institute of Agricultural Research (INRA), was discovering that pesticides and fungicides were responsible for insect outbreaks. His work has largely been ignored. He hypothesised, that an insect would starve on a healthy plant, a phenomenon he termed as 'Trophobiosis.' His book was published in 1985 and finally translated into English 20 years later under the title, *Healthy Crops: A new agricultural revolution.* Chaboussou's theory was that insects don't attack all plants; it is the weakened plants with high amino acids and incomplete sugars that draw in pests like moths to a flame.

In Hawke's Bay, New Zealand, the orchardist Nick Pattison can attest that after removing the pesticide Tokuthion from his programme, mealy bug numbers reduced in the first year, and the next year...they were totally gone. The pesticide was creating the conditions for the pests. In response to human health and environmental concerns around chemical use in the late 1990s, the New Zealand horticulture sector introduced Integrated Pest Management (IPM) strategies, which included hormone disruptors, pollinator strips, improved water management and accurate monitoring. One of the most effective strategies: stop using chemical pesticides! Growers became increasingly aware, that the insect pests were being attracted to disruptions in the trees. Why this information backed by measurable experience, did not flow outwards to other production sectors is baffling.

To address the concerns of growers around increasing pest resistance, many of these chemicals are now being used together to increase their efficacy, which also increases their harm to non-target insects due to synergistic effects in the environment. Research in the past decade, has been unveiling the insidious nature of even low concentrations of pesticides and fungicides on the environment, wildlife, bees, butterflies and on people.[116] [117]

How did soluble, persistent, broad-spectrum pesticides, pass reviews to be released with such gusto into the global environment? It could be argued that these pesticides did not follow a rigorous risk assessment process before being released. "Your risk assessment is only as strong as the question you ask," says Jonathan Lundgren, the agroecologist and entomologist who led the 2018 Regenerative Ag study. He began his exploration into the adverse ecological impacts from pesticides in the late 90s. His doctoral research became more complex and he began to realise that his inquiry into robust risk assessment processes opened a doorway he couldn't close again. "I don't think we can assess risk; the question is just too complicated. The effects are too broad. We don't know which organisms are affected and in what way. How do you do science on 20,000 formulations? As soon as you add an adjuvant, the risk profile changes." The risk assessment process has very little relevance to what happens outside of the lab. In the assessment process around the BT crops, no one asked the question "what would happen if all farmers changed to grow just one or two crops?" Wholescale biodiversity collapse is the answer; above and below ground.

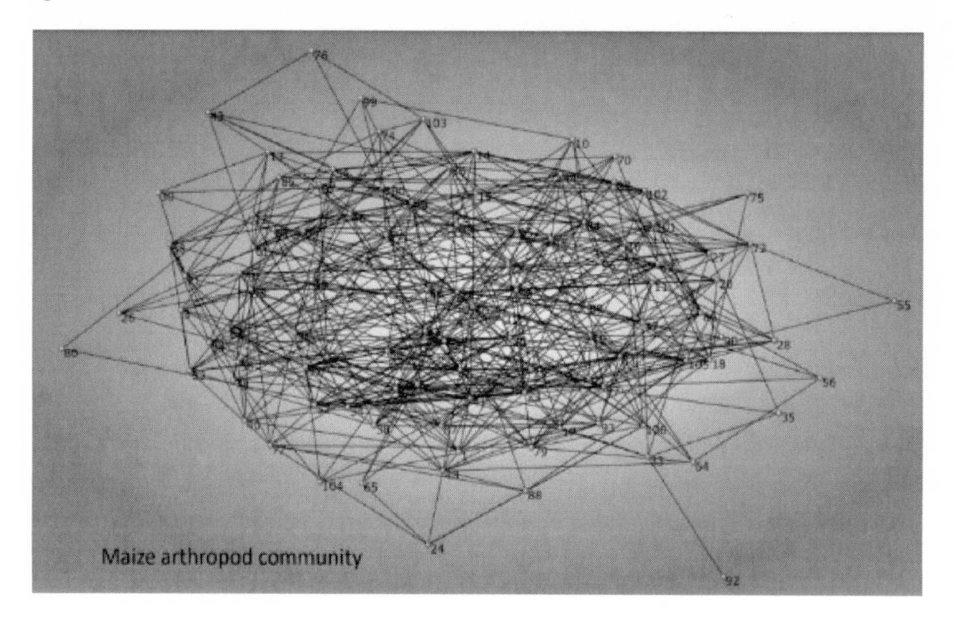

Image: Maize arthropod community networks. Jonathan Lundgren 2011

This image, which at first glance appears to be a toddler's scribble on a living room wall, represents a network of the community connections for insects in maize crops in South Dakota. Capturing data on all these potential interactions would be an ambitious task indeed!

Our scientific questioning does not currently have the scope to capture all that is possible and imaginable from such connections. Scientific research seeks to simplify questions and ways to quantify data and by its very nature must ignore the complexity, synergies and unintended consequences too vast to capture and prove.

Spending time with Dr. Lundgren in the field is like letting a child loose in a candy store. His enthusiasm is contagious. Even the most bug-averse adult finds themselves on all fours, scratching away leaf litter to reveal a treasure trove of newly appreciated 6-legged friends. Lundgren is a world-renowned scientist with over 260 published papers. In 2012, he received the USDA award for "Outstanding Early Career Research Scientist" and directed his own lab. While working for the USDA, he was one of the researchers, who early on, identified the link between neonics and the population decline of butterflies and bees.

Lundgren's focus on pesticides is what he believes led to disciplinary actions aimed at suppressing his work and preventing him from sharing his findings with the public. In 2015, he was part of a lawsuit filed against the USDA, for ordering scientists to retract studies and water down their findings. It takes a courageous and committed person, to be willing to sue a government! It also takes courage for the regenerators and those willing to reject the chemical juggernaut, when it has the strength of funding and the sway to continue to push a failing system.

Observant producers and crop scouts will tell you there is a clear relationship between stress and disease and insect pressures. These stresses may be driven by soil and climatic factors, such as low soil carbon, flooding or heat extremes, soil mineral imbalances or high soluble fertiliser applications.[118] Planting a crop in an unsuitable ecotype is stressful; expecting Summer crops to do well in Winter or vice versa, is asking for trouble. Pesticides also reduce plant Brix levels, change nitrogen dynamics and increase free amino acids. There is a body of scientific literature that recognizes that there is a trade-off between pest insect outbreaks and nitrogen applications. For example, it is "widely established that plant-mineral nutrition is an important determinant of herbivore development."[119] It's curious that these researchers routinely use terms around their findings about insect pests and nitrogen, such as "recognised" and "widely established" when this is not the story being told to producers.

There was a theory swirling among Regenerative advocates, that insects, without a pancreas or method to digest complete proteins, would die after consuming nutrient dense foods. Research by Utah Professor Larry Phelan and others has shown this is not entirely accurate.[120] Their work shows that insects prefer to consume simpler amino acids, which are easier to process than structural sugars or protein. When plants have mineral imbalances, the amount of free amino acids can increase 10-fold. Dr Phelan's work revealed that minerally imbalanced plants produce less proteinase inhibitors, an enzyme plants use to defend themselves against insect pests. His research also showed that when pest insects are fed nutritionally balanced plants, those pests grew stunted, lay fewer eggs and their larvae were less likely to survive. The amount of plant material of the crop that the insects consumed, also reduced. They also discovered that maintaining a ratio of 2:2:1 between B:Fe:Zn, stunted insect growth and reduced free amino acids in hydroponic conditions. Chaboussou was correct: insects are not attracted to healthy plants.

Insecticides, fungicides and herbicides such as Captan, Maneb, chlorpyrifos and paraquat impact directly or indirectly on photosynthesis and on the root exudates necessary for quorum signaling. Enzymes and signals involved in nitrogen fixation and transforming free amino acids into complete proteins, are compromised by these chemicals.[121] Again, a perfect sales pitch for someone not the farmer.

Plant photosynthesis drives everything. As plant health and Brix levels increase, pest pressures decrease. A study carried out by a French student, looking at data taken from over 22 regenerative properties in New Zealand, showed that a Brix level of 12 or above for grasses and 14 in legumes, throughout the growing season, was effective in reducing pest pressure. As a direct result from their soil and pasture health programmes, these farmers saw a lift in Brix levels, while their insect pest pressures dropped. Insects are not after every blade of grass or every apple. They're looking to clean up garbage!

Establishing and maintaining a diverse microbial community and building soil organic matter, is essential to reduce environmental stressors, such as drought and temperature fluctuations. Microbes can reduce nutritional stresses by ensuring plants receive minerals, at the rate, time and form they require. Soil and leaf microbes provide a range of beneficial services to boost plant health. Many of these effects can be indirect. Take mycorrhizal fungi, for example; they can increase plants' ability to emit the volatile organic compounds (VOC) that attract insects. These VOCs attract beneficial predatory and parasitic insects to target the pests. In the presence of mycorrhizae, plants emit VOCs which boost predatory mite populations by stimulating egg laying and shortening egg maturity time. [122]

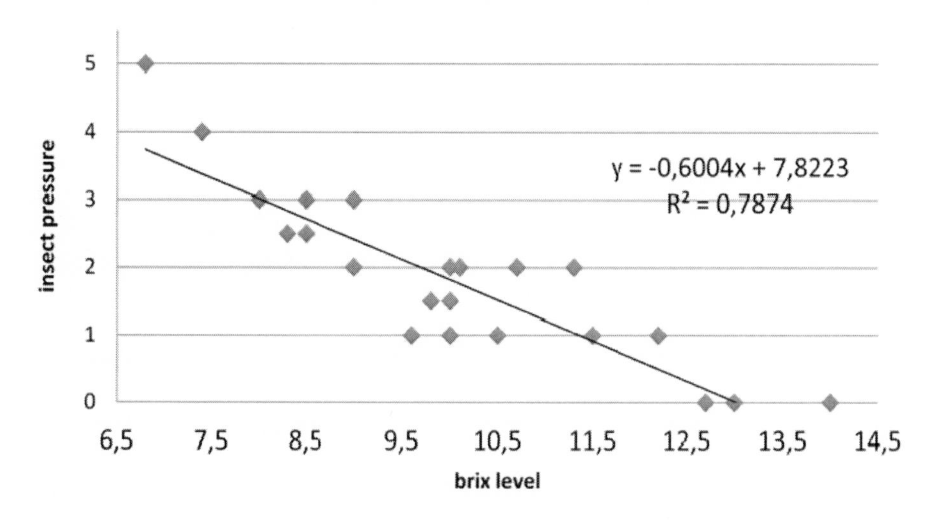

Correlation between Brix and insect pressures from 22 properties in the Hawke's Bay and Manuwatu regions in New Zealand. Florent 2001[123]

Perhaps it's becoming clearer, that there's a gap between the soil health you're committed to having and its current state. What to do next? Swinging from one branch of management to the next, can be tricky if you don't want to risk yield losses from high disease or pest pressures. On the Lindsay racing bloodstock farm, there were major factors hindering progress. With compromised minerals and biology, the plants were under stress, and insect pests were having a literal field day! Between the crickets, mealy bugs, grass grubs, clover weevils and Lucerne (alfalfa) fleas, the insect pest biomass far outweighed the mass of horses above. Something needed to be done so their soil could take a leap forward.

As insects prefer to feed on low Brix, high amino acid (N) plants with an imbalance in minerals, the low soil carbon, compacted and shallow soil conditions at Lindsay racing were rolling out the welcome mat. Tight compacted soils offer ideal conditions for many insect pests, with poor nutrient mobility and bacterial boom and bust cycles compounding the problem. Compacted soils offer a sanctuary for insect pests as their natural insect disease controls, the entomopathagenic (EP) organisms, are impaired in low oxygen. As part of a transition programme, EP fungi, bacteria, protozoa or nematodes are excellent interim tools, if insect pests are putting pressure on valuable production. These microbes exist naturally in healthy balanced soils and inside plants across the world. To thrive and reproduce, they need the bodies of insects to be available. These organisms can have dramatic and significant impact on pest populations. A Cornell study using EP nematodes on the destructive weevil, plum curculio, reduced pests by 70-90%.[124] EP have other plant health benefits; providing nutrients directly to plants, defending roots and increasing plant immunity to disease, known as induced systemic resistance (ISR).[125]

With the high insect pest pressure at Lindsay Farm, we applied a proprietary blend of 4 different EP fungi and bacteria with a fish hydrolysate and saw a dramatic drop in the presence of clover flea, weevil and mealy bugs. The grass cover and root depths increased, grass pull stopped and the proportion of clovers increased. This application enabled the soil building processes to really kick off, as the bugs chewing on root systems were eliminated. We also applied *Metarhyzium* and *Beauvaria* strains for ticks, spraying them onto the soil and applying them along the horses' back line with some oil as a carrier. As a result, there was a massive reduction in the tick population. Researchers have found these organisms can control ticks by 80-100% in the field.[126]

You can apply packaged bio-control products. However, we've been involved in multiple projects now and have witnessed again and again, that as soil health improves, these bio-controls naturally show up and thrive.

How can you tell? Start by digging and look for mummified bugs. For instance, the presence of three common EP fungi can be identified by different coloured fuzzes of mummified bugs: green, white or yellow. The white fuzz may be *Beauveria bassiana* (white muscadine disease); the green may be *Metarhizium anisopliae* (green muscadine); and the yellow fuzz may be *Isaria fumosoroseus* (aptly named yellow muscadine). The yellow muscadine disease is a pathogen to over 25 different insect families, including the diamondback moth, Russian wheat aphid and silverleaf whitefly. *Metarhyzium anisopliae* infects over 200 insect pest species, including termites, thrips and, along with *Beauvaria bassiana,* is being researched as a bio-control agent against the mosquitos that carry malaria. *Heterorhabditis bacteriophora* is an EP nematode, now commercially available, which loads up their hosts with bacteria and eggs. This biocontrol is delivering great results in controlling fleas, ants, leatherjackets and weevils. As their young hatch, the baby nematodes eat their insect hosts from the inside out. In my analogy of soil as a horror movie, these are the aliens popping out of your chest!

Researchers around the world, have struggled to get consistent results with EP strains and, to my mind, many good projects have been shelved too early. In the field, we have consistently used fish hydrolysate and a humic acid to deliver EP organisms to the plant. Using these inputs is key to the survival of the inoculants and ensuring delivery into the plant. Many insect pathogens are also endophytes, meaning they live in harmony with plant tissue without harming their valuable host.[127] Many plant species naturally harbour these little stealth bombers, including and not limited to: artichokes, bananas, cocoa, beans, cotton, corn, oilseed rape, coffee, date palms, opium poppies, pine trees, pumpkins, sorghum, tomatoes and wheat.[128]

There is now mounting evidence to support this, but through field observations with farmers using these products, we saw how quickly the fungi worked. Within days, following a single foliar application, the fungi travelled through the xylem to the roots. Here, they could directly target any insects chewing on roots deep into the soil. Both *Beauvaria* and *Metarhyzium* spp are endophytic EP fungi. Like mycorrhizae, they live in and on plant roots. Until recently, scientists did not believe these fungi provided any other services back to the plant. Now researchers are finding, not only are they effective insect-killing machines, they also increase iron uptake and deliver valuable nitrogen from the insects' bodies back to the host plant! In forest ecosystems, 70% of the N returned to soil, comes from dead bugs or their poop. They are a significant contributor to the global N cycle (and ignored in all N-cycle teachings.) Across a hectare of soil, there may be as many as 172 million insects (72 mil/ac) weighing around 450 kg. As insects are the nitrogen thieves in any ecosystem, in vibrant soils, they may be providing an additional 45 kg N/ha (40lbs/ac) from their cold dead bodies.

As many of these fungi can be non-specific, careful consideration is required before application. The jury is still out on the potential impact from *Beauvaria bassiana* on dung beetle populations in the field. Researchers in New Zealand, have seen this disease infect old and injured dung beetles. *Metaryzium* itself is safe for bees and is being promoted by the passionate Dr. Paul Stamets as a mycelial approach to the control of sudden colony collapse and varroa mite.[129] Watch Paul Stamets' TED talk on "How fungi will save the world." Actually, while you're there, watch his entire video library. Fascinating stuff from a fascinating guy. Our anecdotal in-field observations, have seen increases in incidence of beneficial insects, including dung beetles after applications of *Beauvaria bassiana.*

The effects on different properties, have at times been extraordinary and, at other times, total failures. Failures, although painful, are good learning tools. Our failures have built our knowledge bank so we're clearer that getting timing right, is super important in targeting specific insects.

Bio-pesticides must be applied at the right life stages; when the insect pests are actively feeding on plants and their roots. You need to use the right fungi/bacteria in a blend. Many EP are UV sensitive and it's preferable to spray after the bees have gone to bed or at least use strains which will not hurt bees. Actually, there are so many other vital pollinators besides bees; just spray at night. Once. Don't worry about what your neighbours think; they already think you're crazy! We still err on the side of caution of maintaining ecosystem balance and only use these bio-controls in the early transition stages.

A few years ago, I met a dairy farmer in North Canterbury, New Zealand, who had been having issues with slow pasture recovery and grass pull; literally the whole plant pulls out when grazed. Hauling out the trusty shovel revealed an army of fluffy white spots throughout the soil, which were creating root rot. There are some consultants who have been misidentifying this pest as beneficial fungi; sadly, it is a pasture pest called "mealy bug" and, in my experience, it appears to be on the rise in many Ag regions. They are found in more temperate, warm and moist environments, causing trouble in turf grasses and high production crops in Florida, Texas, South-East Asia, Eastern Australia, New Zealand and South Africa. It is being attributed to the disease called "pasture die back", which is decimating Australian grasslands.

Dig a hole; you may be shocked to find that you have more herbivorous livestock under the grass than above.

Looked at through a 20 X magnifying glass, adult pasture mealy bugs appear headless and legless—pinkish, oval-shaped insects up to 2mm long. They look similar to the mealy bugs you'll find on sick house plants. They are often found in the crown of the plant, with the females surrounded in tufts of white, waxy secretions. They appear to take a fancy to clover and suck the sap from grass roots. A heavy infestation can kill plants, with effects often more obvious in Autumn. While they're sucking away, they're also producing hydrophobic waxy exudates. This creates another vicious soil cycle, contributing to water repellent soils.

This dairy farm had a massive infestation. In the long run, he needed to address his poor soil health and compaction; in the short-term, I recommended he order some bags of *Beauvaria*. He applied the product and the pasture pulling stopped within a few weeks. We dug holes across the length of the farm and didn't see a single mealy bug, until we reached the last paddock, which was heaving with white fluff. "Ahhhh," he said. "So *that's* why there's a bag left in the shed!" He had accidently run a very good trial, which helped him see that the results weren't just a coincidence or due to seasonal changes.

As a reminder, these natural biopesticides are in healthy soil and plant environments. Build soil health and they will come! Applying a biopesticide is not a transformation in thinking; it is a substitution. Applying any of these bio-controls is like keeping an aircraft speeding along a runway. They provide one tool for us to move quickly along our journey; however, at some point you need the plane to lift off!

Diversity is key in reducing insect pest pressure, making it harder for them to find food and providing habitat for the predators. In heavy infestations of root feeding organisms, crop rotations are essential to break the insect lifecycles. I find it fascinating how pests in the Corn belt, have evolved to have a 2-year cycle, to miss the intermediate Soy rotation and coincide hatching with their favoured foods. Mixing up what species will be grown and planting crops as polycultures or multi-species, reduces the build-up of pests. This diversity provides a broad array of quorum signals, pest deterrent compounds and hosts for EP. Jonathan Lundgren is clear; the root cause of pest population explosion is the encouragement and expansion of monocultures. Management has been the driving issue and diversity the key to any successful pest management strategy. Will we ever exterminate a pest off the planet? No, and I don't think that's helpful thinking, in a holistic system, where everything has its place.

A recipe for poor soil health also provides the ingredients for disease: compaction, waterlogging, soluble nitrogen, low carbon and monocultures are all contributing factors. Many of the drivers that give rise to weeds and pests, are the same drivers for disease. There are practical ways to proactively reduce diseases; including grazing techniques, hygiene clean-up techniques and recipes, cover/intercropping and avoiding tillage etc.

If diseases are already present, causing critical production losses (particularly in high value crops), then we adopt a 4-pronged approach. This includes actions which aim to control the disease using biologicals and nutrition, reduce the spread of the vector, increase plant resistance and health and ultimately address the underpinning structures that created the conditions for disease in the first place.

Actions to reduce infection depend upon the crop type, disease species present and management options. Does an organism spread through water, wind, machinery, or livestock? Does it overwinter on debris? And what climatic conditions does it require to spread? Asking these questions will help you to act to reduce further spread. Typically, with cropping, gardening and horticulture businesses, actions need to be taken to decompose organic materials and reduce disease build-up between growing seasons. There are commercial digestion products available, or you can make your own, using molasses, fish and calcium products.

Disease organisms are present in every soil environment. Infection is a sign that something else has gone awry. Most diseases are 'secondary vectors,' such as the root rot, phytophera, which target dead materials. Another factor has created the environment for rots to set in. Recent studies concluded that it is a deficiency in defensive microbial partners, which invites diseases in.[130] More research is desperately needed into what the determining factors behind diseases are; is it a lack of microbial defences, key minerals or management?

Simon Osborne, the regenerative cropper in Canterbury, New Zealand, has been trialing and measuring changes to his crops for decades. He is a great example of putting the principle "measurement to inform management" into practice. Over many years, using Conservation Agriculture (CA) practices, the farm system has steadily evolved as soil carbon has increased.

However, the system continued to rely heavily on herbicides, fungicides and chemical seed treatments.

By 2016, Simon felt the farm had plateaued and, under this farming model, "no further progress was being made in soil health and overall farm ecology although financial performance was great." Feeling demoralised with what he calls "prescriptive chemotherapy farming," the family seriously considered selling up and doing something else, until an off-chance discovery of the book 'Teaming with Microbes' by Jeff Lowenfels. The book opened a doorway to "the beauty of soil ecology and the basic principles of a synergistic biome." Simon has an active mind, between reading technical literature and attending workshops by regenerative educators, he had the confidence to start his transition to a low-input ecological system.

Simon found reducing his chemical fertiliser use was easy, as he had "already worked out that less is more." In fact, he discovered "none is more in many cases", as he began to realise fertiliser use was increasing both costs and insect and disease pressures. The biggest challenge was the elimination or reduction of chemical 'icides, detrimental to plant physiology and the soil biome. High value seed crops like hybrid radish, are managed intensively with cultivation and a heavy calendar-based chemical regime; up to 50 hits of fungicides and insecticides can be applied during a single growing season. This is a big treadmill to step off.

Simon is always running multiple trials, particularly when looking to cut fungicides from his valuable seed crops. With chemical inputs, costing seed producers well over NZ$4,000/ha (US$2700), potential crop losses put producers into a risky position. Profitably eliminating fungicides and pesticides, has been a focus and a goal for Simon. He became increasingly aware, that unless he removed these chemical inputs, his soil health programme could not progress. His observations and testing across multiple crops, showed that all systemic chemicals were 'terminal' to microbiology. (The only exception Simon has found, is low rates of Triazole used to control rust in grasses. Unfortunately, this is a fungicide known to harm bees.) Using his Brix meter in a side-by-side trial, he recorded sudden drops in Brix levels following fungicide applications. Crops without fungicide treatments looked far healthier, with a deep green colour and more even-sized seed production. After applying selective herbicide applications, Simon also measured a decline in calcium and phosphorus in his crop. He was initially wary about sharing his experiments with others. One day, when his seed buyer was viewing the crop, the buyer commented that it was looking so good, they wouldn't require any further fungicide treatments, which led Simon to finally reveal that the crop had in fact been spray-free all along.

Research has shown that fungicides have a varying impact on plants and the microbiome. Many disrupt a broad array of physiological functions in plants, from a decline in photosynthesis, to reductions in root mass.[131] Many fungicides have broad non-target effects on soil microbiology, mycorrhizal hyphae and beneficial insects.[132] Fungicides are detrimental to regenerative soil health goals. Pre-emptively pulling these inputs is not a sensible step without using transition tools, unless you can afford the potential yield drops.

As Gary Zimmer, the biological agronomist, states "you need to earn the right" to pull these chemical props. This can be done in the short term, through substituting with nutritional and biological supports. On those operations that can risk not using chemical controls, you need to be willing to go through a potential 'ugly hair stage' (you girls know what I mean!). There may be cosmetic damage, which can be hard to watch when coming from a high-input chemical system, but this visible damage often results in little to no loss in production and will certainly save you some money.

Once while on a field trip with Di and Ian Haggerty in Western Australia, we looked at a crop in its first-year transitioning into their low-input biological programme. The previous owners had thrown an entire kitchen sink of inputs, with multiple herbicide and fungicide applications. The crop was in poor shape with net blotch and yellowing evident. The conventional agronomists on the tour looked uncomfortable, frowning at what they saw as a total disaster. They commented that the crop desperately needed fungicides and nitrogen. Instead the crop received worm juice and compost extracts.

I wish these agronomists had stuck around for harvest. This ugly little crop yielded the same as their neighbours, at far less cost. The following season, the area was hit by catastrophic frosts, with some producers losing up to 80% of their crop. The ugly duckling block lost only 5%. The cost of having a poorly looking crop quickly paid off major dividends and fast-tracked their soil health programme. It is key to stop blowing the microbial bridge.

In healthy soils, there are complex dynamics at play, that enhance a plant's ability to protect itself against pathogens and parasites; the induced systemic resistance (ISR). When plants come under attack, they set off a complex communication network, signaling for support from specific beneficial microbes with a variety of biochemicals. Organisms, such as *trichoderma* or *bacillus*, respond and provide the trigger to activate plant defence pathways. This is under the proviso that plants are healthy, have optimal Brix to feed microbes and the microbial diversity is present to respond.

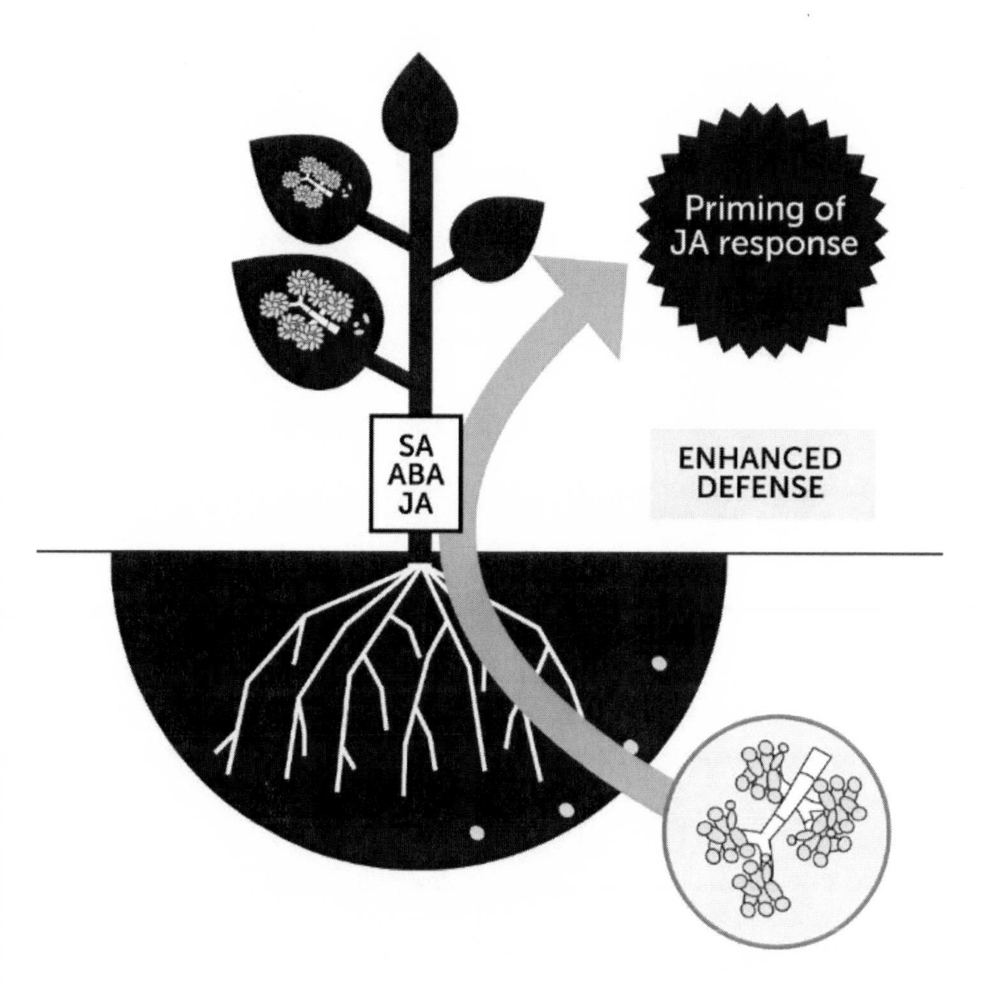

Image: *The plant signals to the trichoderma fungi (on right), to provide the trigger. The fungi release an oxidative burst which sets off a cascade of defence network: jasmonic (JA), abscisic (ABA) and salicylic acids (SA). Without these microbial relationships, the defence network is disabled and diseases, such as botrytis, run rampant (left side leaves in image). Image adapted from* Martínez-Medina, et al (2013).[133]

There is increasing interest in the use of these plant defence elicitors. Some of these elicitors can cost-effectively "prime" the plant, much like a SWAT team on call, providing increased resistance and ready to provide a rapid response to attack. Research has shown a significant reduction in pests (spider mites, caterpillars and aphids), as well as diseases (such as botrytis and powdery mildew), following applications of jasmonic (JA) and salicylic acids (SA). The effects from seed treatments with these elicitors, can last far longer than the chemical prophylactics. Plants treated with elicitors were also better equipped to switch on their own defence genes when under attack, versus the use of chemical pesticides, that suppress their natural defences.[134]

In Simon's soil health programme, bio-control agents are applied as seed dressings and foliar applications, to ensure valuable plants are protected in the transition. These bio-controls include *Trichoderma viride, Bacillus subtilus and Streptomyces lydicus* for disease prevention. Trichoderma is a fungi-eating fungus that lives symbiotically on plant roots. It is used commercially to control fungal diseases in a wide range of crops. Simon has also applied seed dressings containing fungal EP for insect pest control, as well as diverse mycorrhizal inoculants.

The future for fungicides and pesticides lies with biomimicry. Plants rely on a complex system of biochemical and electrical signals that provide a multifaceted surveillance, defence and detoxification system. They produce a variety of elicitors, proteins (peptides), enzymes and phyto-hormones, to defend against pathogens and other invaders.

When triggered by wounding, pathogens or insect attack, elicitors are triggered. These activate the plant's chemical defence system. Proteins include ribosome-inactivating proteins, protease inhibitors, lectins and anti-fungal peptides, that represent the first line of defence in both plants and animals.

Elicitors include benzoic acid, methyl salicylate, chitosan and plant hormones such as auxins, gibberellins, strigolactone, cytokinins, abscisic acid, ethylene and jasmonic, salicylic and b-aminobutyric acids. Functioning much like our own hormonal systems, there is a cascade of interactions. Together these bio-chemicals orchestrate a complex defence system:

The phytohormones (auxins, gibberlins and cytokinins), have long been recognized for their roles in plant growth and development. They also have roles in the complex cross talk with other plant hormones, when plants are under stress. More recently, strigolactones (SL) have received increased scientific scrutiny.

Phytohormones are exuded out roots, promoting symbiotic interactions between plants and soil microbes, including initiating rhizobia and mycorrhizal symbiosis, growth and spore germination. They are involved in the priming of plant defences.

- *Jasmonic acids (JA) are produced when a plant comes under attack. When activated, JA controls the synthesis of defences, that increase plant resistance to pests and pathogens. If an insect does get past a plant's defence system, JA then degrades the essential amino acids that insects need for digestion. When the JA pathway is initiated, the plant sends signals and carbon to the rhizosphere to stimulate disease suppressing bacteria.*

- *Salicylic acid (SA) plays a role in plant growth and development. SA is essential for plant immune signaling and a regulator for many plant defence and stress responses.*

- *Abscisic acid (ABA) works together with SA, to play a pivotal role in reducing stress and increasing plant defence against pests and disease.*

- *Ethylene (ET) regulates the genes involved with plant defence, switching them on during times of attack. Ethylene works together with the other phytohormones.*

- *Chitinase is an enzyme that breaks down the chitin bond present in fungi and insects.*

- *B-aminobutryric acid (BABA) is an important hormone that primes the plant to quickly and efficiently respond to stress and attack. These plant-priming effects can be long lasting and can be passed on as a type of memory, informing future generations about potential stressors and diseases.*

- *Glutamate is an amino acid that acts as a neurotransmitter to trigger rapid long distance signaling using calcium.*

Protection from induced systemic resistance (ISR) has been reported against a wide range of pathogens, including systemic viruses such as cucumber mosaic virus, root-knot nematodes, leaf-spotting fungal and bacterial pathogens, sclerotinia, crown-rot, stem-blight fungal, damping-off, take-all and late blight diseases. In most cases the microbes that stimulate ISR also promote plant growth.

Organisms known to support ISR include Pseudomonas spp., Trichoderma spp., Mycorrhizae spp., yeasts, Rhizobacteria, Bacillus spp. and other gram-negative bacteria.

Kelp and green algae also stimulate this plant defence system.

Bio-chemical elicitors offer a powerful interim tool; they do not carry the same risk of resistance as chemical controls due to their diverse modes of action.

Minerals also play an essential and vital role in plant defence. These include calcium, silica, zinc, copper, manganese and other trace elements. Simon Osborne tests plant tissues during the season and applies carbon-based foliar minerals as needed in the growing season, to reduce plant susceptibility to pests and diseases.

Calcium improves the environment for beneficial soil microorganisms. It is essential for all plant growth points and a key component of cell walls. Any deficiencies in calcium, therefore, create vulnerability to infection. Calcium has a major role in plant defence and has been shown to inhibit fungal spore germination. For tree crops, it is critical to mobilise calcium after harvest, to set trees up for bud burst the following season. Calcium availability to the plant is limited by low boron levels, poor biological activity and soil moisture deficits. As fungi play an essential role in making calcium available to plants, poor soil fungal activity leads to an increase in disease.

Milk, ideally raw, or even powdered, can be a great foliar application. It has numerous beneficial properties for soil and plant growth, including beneficial microbes and bio-available calcium. It contains proteins and other compounds which have been observed to suppress plant disease and enhance plant tolerance to heat stress and nutrient uptake capabilities. Furthermore, many of the bacteria ubiquitous in raw milk, are known to be beneficial plant growth promoting soil microbes. Ferroglobulin, a protein in whey, produces an oxygen radical when exposed to UV light, which may be toxic to fungal spores.[135]

As the second most abundant element in the Earth's crust and the soil, silica (Si) has been largely ignored by agronomists. It is not considered an essential plant growth element. Silica is crucial, however, as it provides plant defence against pests and fungal/bacterial disease and reduces plant stress. It is a cell-strengthener and an activator for many plant functions. It also lifts low electrical conductivity (EC) in soil. Research has shown it has an essential role in reducing lodging, drought, salinity, excess nitrogen and phosphorus and heavy metals. There are many factors which reduce Si uptake to the plant: compaction, lack of air and water and low organic matter. To address such issues in rice paddy fields, Si is routinely applied. Silica reduces pest and disease effects by strengthening cell walls, assisting plant chemical defence and ISR. The addition of potassium silicate, for instance, substantially increased the effectiveness of *Beauvaria bassiana* in pest mite control.[136]

Copper has been used as a fungicide since the late 19th century in French vineyards in the well-known Bordeaux mixture. Copper is essential for plant health, growth, development and protein formation. It directly inhibits fungal spore development and has a role in regulating ISR defence responses in plants. Like all trace elements, copper can be a biocide, so teeny tiny amounts are all that is needed. Copper is used to provide plant nutritional needs, not to kill things.

Cutting out fungicides and pesticides can feel like the riskiest step for many producers, but this is often the first stage many successful producers like Di and Ian Haggerty or the Twin Rivers operation take, in their journey towards achieving soil health. Australian cropper Grant Sims, dropped his pesticide use in 2008, replacing them with nutritional and biological inputs across 3,500 Ha (8,650 ac). Despite the dry conditions he was operating in, he saw immediate results with a rapid increase in his underground workforce, including fungi and earthworms. Above-ground, the decomposition of crop residues was kick-started, and plant health responded to the improved nutrient cycles.

A trip to the Sims farm in Northern Victoria, is a staggering experience, demonstrating the potential for cropping systems across the world. In this harsh dry environment, we took a penetrometer into the field and pushed it into the ground. It drove down through the soil, like a hot knife through butter. As a comparison, across a fenceline, the penetrometer stopped at an inch. For the past 5 years, the farm has received half its average rainfall, with only 220mm (8.6") soaking into the ground and, yes, when you ask Grant how much rain he gets, the answer is, "all of it."

Grant is an innovative, creative individual, who has developed his own equipment, compost extractors and liquid mixing systems. Through supporting plants nutritionally and addressing compaction with calcium, controlled traffic and zero tillage, he has eliminated the use of pesticides, fungicides and chemical seed treatments. By blending nitrogen with a carbon source, he uses only 25% of the total nitrogen typically recommended. With all the reductions in inputs and costs, there has been no negative effect on production. As a result, the farm has become far more profitable. According to Grant, "The key is getting the soil chemistry in balance and building up residue to provide food and a good home for the biology to grow. Get that right and everything else—the profits, the crop, the yields, gross margins—will follow."

He manages for maximum plant residue, using a stripper header to leave a deep base of stubble and litter to build organic matter. A 'controlled traffic' approach is used to keep machinery confined to narrow tracks, reducing field compaction. The focus on soil health and reducing chemical inputs, has created more enterprise opportunities for the family. Seeing the value from adding livestock back into a cropping system, he's also looking to expand a 'paddock-to-plate' beef business.

Grant and his wife Naomi are also setting up a cover crop seed business. Cover crops are a key tool in feeding microbes, building organic matter and soil cover; all of which add to the beneficial signaling processes. These crops lay down an armour to protect soil and provide habitat for beneficial insects. There has also been a focus on providing habitat for wildlife, reconnecting corridors of native trees and shrubs and encouraging natural vegetation around the edges of crops. Entomologist Jonathan Lundgren visited the farm in 2016. By all accounts, he was even more excited than usual, after discovering a huge diversity of insects not typically found in cropping lands.

Successful regenerators fast-track their transition off biocides by loosening the reins on what people think a crop should look like. It's important to keep your focus on the long-term picture and avoid worrying about trivial details. As Glenn Stahl, from Twin Rivers told me, "I'd rather have a messy farm, than a messed-up one." It certainly takes something, to allow a crop to look less than the normal, 'clean' and artificial version, that has become the norm. Tolerating a few weeds, some cosmetic rust and insects chewing a couple of leaves while working on building the microbial bridge, may take a year or two. Use transition tools to ensure you are still harvesting a profitable crop and breathe through this stage.

Chemical controls are designed to keep crops 'looking perfect.' It is this cosmetic, short-term approach, that keeps the growers under agribusiness control, creating more unintended consequences. Instead, through a focus on stimulating plant defences, increasing Brix and microbial signaling, the benefits continue to accrue to the regenerators over the long-term. As Grant has found, striking out into new territory is not always easy, but it often yields the biggest returns.

When Jonathan Lundgren left the USDA, he set up Blue Dasher Farm and Ecdysis Foundation in South Dakota, a demonstration farm for regenerative agriculture. He is now able to make a bigger difference with producers on the ground and provide research, without the risk of being muzzled. Lundgren is clear: "We're using all of these pesticides because we've created a pest problem." Diversifying farm systems above- and below-ground is key to regenerating ecological health and reviving rural communities. "Regenerative agriculture is the future. It solves so many issues, at so many levels."

For all that he has been through, there must be some solace in hearing that in 2019, the EPA has cancelled the registration of 12 of the 59 neonics on the US marketplace.[137] This is one step in the right direction. Now let's work to create resilient healthy ecosystems, that don't require the chemical inputs. Regenerative agriculture offers a total rethink on how healthy crops are grown, one which supports optimal health and biological synergies. It is the only way we can finally put an end to this 'war on nature' and instead find peaceful and real solutions which work with the environment.

Strategies to reduce insect pests

* Diversity, diversity, diversity!
* Cover crops, intercropping, wild field margins, crop rotations, alley crops, mosaic of diverse crops. Feed a diversity of soil microbes and beneficial insects.
* Increase Brix level.
* Reduce soluble nitrogen use.
* Reduce disturbance.
* Reduce free amino acids and plant stress.
* Consider if this is the right crop for my environment?
* Increase your observation skills around natural insect dynamics.
* Opt for biological seed treatments to increase elicitor responses.

Biocontrols in transition

If using bio-pesticides during the transition, use with the following application considerations:

* Evening applications are preferable, as many pollinators are resting and spores are inactivated by sunlight.
* Mix and apply with fish hydrolysate.
* Need to be applied during larval stage of the pest life cycle.

Note: biological inoculums do not tank-mix with fungicides, or trace elements such as copper or zinc. To optimise pest reduction, keep a focus on the soil health principles.

13

The Future is Now

"You don't hope for something that is. It's always future oriented, which means, hope is inherently intolerable of the present. The present is never good enough. Our time requires of us to be hope free. To burn through the false choice between hopeful and hopeless... it's the same con job." - Stephen Jenkinson

Every moment of time I spend with Dr. Patterson Stark, I'm convinced, returns me a tenfold of life on this planet. Dr. Stark is a practitioner of Lifestyle Medicine and a billboard advertisement for the benefits of walking your walk—a living, breathing, mountain biking, surfing and meditating demonstration of the benefits from applying Lifestyle Medicine to a person in their late-60's. He balances his demanding work schedule, by ensuring he takes time to do the things he loves, which, as I discovered on one occasion, includes jumping off mountains in his paraglider.

Dr. Stark's journey began in 1986, with a terminal cancer prognosis. When he was told "You have two weeks to live," he followed the conventional recommendations and cheated the reaper. To ensure cancer never reared its ugly head again, he dug deep into the research. With his findings, he altered his diet and lifestyle, before moving from Colorado USA, to Christchurch, New Zealand. Dr. Stark's approach reflects my approach to working with soil and landscapes. He asks questions about history, diet, exercise, stress and 'work-life balance.' He diagnosed an insightful cascading effect, between foods, detoxification and hormonal imbalances which had been plaguing me since my incident with paraquat.

His research and over 30 years from running a successful practice, revealed a simple truth: lifestyle choices are key to preventing and reversing the barrage of health conditions facing the developed world. As much as 97% of conditions, such as obesity, diabetes, sexual dysfunction, metabolic heart diseases and cancers, can be addressed through diet, exercise and lifestyle. "Just diet alone" he explains, "can reduce the risk of degenerative diseases by 80%." This is not new information to humankind; the Egyptians knew this, writing an encryption in a pyramid 3,800 BC ago: "Humans live on one-quarter of what they eat; on the other three quarters lived their doctor."[138] The Lifestyle Medicine practitioners, take a holistic look at underlying drivers to health, which cover: "physical inactivity, nutrition, smoking, alcohol overconsumption, chronic stress, sleep deprivation, social isolation, loss of culture and identity, exposures to toxins and other influences of society and environment."

Currently, most doctors, like agronomists, are not educated about nutrition; instead, they focus on treating symptoms. The success of Dr. Stark's work and others has led to the development of a curriculum to train front-line medical doctors in preventative medicine. Training is now available in 108 medical schools in the U.S. Thirty-seven hospitals, including government hospitals, are rolling out programmes that include switching to a plant-based diet, no oils, meditation and movement. Don't fear, my carnivorous friends; plant-based doesn't have to mean zero meat!

I personally find from living on the road, trying to source healthy, nutritious foods, offers unique challenges, particularly in the U.S. To support my own health, I have been taking quality supplements. These supplements only provide a weak facsimile to the potentially 1,200 different phytochemicals, vitamins and enzymes that a plant-based, nutrient-dense diet can provide. Dr. Stark clarifies that when people choose to take a supplement like vitamin E, there may be one or two types of the vitamin in a pill, while there is potentially 10 or more in a plant. Five walnuts for instance, will give you all the vitamin E that you need for protection and this amount "is shown to stop brain atrophy." Or take the bioflavonoids of vitamin C. "What kind?" asks Dr. Stark. Most people he sees, are taking a processed and sterilised version, which was probably made from GMO corn. How many bioflavonoids are there? "Potentially over 1400."

It doesn't take a degree in rocket science to connect the dots for human health; we require whole, diverse and complete nutrients and vitamins, low in heavy metals and toxins, from food. The quality of these foods intrinsically relies upon the organic matter, biology and mineral function that food was grown upon. In this way, our health and wellbeing is all indirectly or directly related to microbial activity in soils and the extraction of plant medicines. Junk food in, is junk food out. I've been in three science meetings now with respected agricultural scientists. They have argued, somewhat heatedly at times, that there is no link between how food is grown and human health. It can't be related, as "we're living longer."

Human populations in underdeveloped countries don't see the same health conditions experience in the developed world. In our modern world, health is framed as the absence of disease; a view which is also crudely reflected in soil health. To be truly healthy, systems need to be vibrant and functional, not propped up by artificial supports. In developed countries, more than half of the population are on more than 2 medications by the age of 65. The sad state of the modern world is that we're not living longer, we're dying longer. Dr Stark is adamant; "If we continue on this trajectory, we will go bankrupt. We won't be able to continue to do 500,000 coronary bypasses year after year, it's not working." Just by addressing lifestyle, Dr. Stark calculates we can cut annual health costs which exceed $50 billion to less than $5 billion.

Eating a diverse diet, keeping at least half our plates covered in plant-based materials, is one place to start. Taking on a simple challenge to eat at least 30 different foods a week, is a revealing experience. Michael Pollan's excellent book, The Omnivore's Dilemma, shows how the American diet could be traced back to one crop: corn. Of the planet's 15,000 edible plants, 95% of the modern Western diet contains less than 30 commercial crops. This has an impact on the environment, biodiversity, resilience and human health. The benefits from 'eating the rainbow' of unadulterated foods provides a multitude of synergistic effects, which are difficult to anticipate and to measure. Dr. Stark gives the example of a patient with cancer. These cells can be put into a petri dish with lycopene and cancer cells reduce by 30%. Take those same cancerous cells and add a drop of lithium and there is no response. "When the two are put together we could expect a 30% reduction of the growth in those cancers, but we don't, we get a 70% net reduction." Just as we find when working with soils, these synergies are extremely difficult to predict.

I've spent a few years feeling exhausted. I was hoping Dr Stark would have a silver bullet or magic pill to lift my energy levels. After a physical exam, talking about diet and lifestyle and looking at my blood work, he offered his prescription: "rest, exercise and more greens." I was a little disheartened. No magic pill? Stop travelling so much?! Then it struck me; I was not being regenerative in my thinking around my own health! If nearly all health issues come back to diet, exercise, lifestyle and mindset, surely this is an easy fix?! These are the habits of a lifetime, however -- ones tied to our identity and the hardest things for us to shift. It's far easier for me to pop a pill, than to cut down on travel, exercise and fill my plate with 50% greens. This is the same with actions on the land. Over the past century, farmers and ranchers have been trained to ask for a prescription, or technological fix for what comes back to common-sense actions: the "diet, exercise and lifestyle" guide for soil.

Although it would be nice, there is no pill, prescription or calendar to regenerate soils, aspects which once made conventional agriculture so simple to adopt. I say 'once' as these simple pills are now becoming increasingly complicated and costly. What was once a single herbicide and 250kg of product, has become 5 herbicides, fungicides and "Have I got the solution for you!" The uncanny parallels between what is happening with soil health and chemical inputs and the pharmaceutical approach to human disease, reveals a deeply ingrained dysfunction. Like patients, producers are being encouraged to think less and ask fewer questions. Regenerative agriculture is the polar opposite of this paradigm. It requires deeper observations and inquiry. It begins with a curious mindset, which in turn informs the questions to ask and the actions to take. For people who are stressed and busy, this can be the most daunting and challenging of steps to take.

The holy grail for healthcare, conservation, landcare organisations or climate change mitigation is the answer to the question: "What does it take to shift human behaviour?" Many of the people interviewed for this book, feel the future looks bleak for agriculture right now—unless people are willing to do the thinking and take the steps required to transform their land. Most believe that changing rural behaviour will require a crisis. The 19-year-old daughter of Roger and Betsy Indreland, Kate, believes it's going to take more "sick kids and doctors coming up empty" or extreme climatic events like drought, to make her community realise there is a need to adapt.

Many regenerators commented that a crisis or more suffering is required to shift behaviours. This idea is partially supported by the rapid growth in interest in many Australian communities where climatic variability and low organic matter is making a risky farm business even riskier. Despite the suffering that is happening in regions experiencing below normal rainfall and debilitating debt, many continue to farm as usual, adding inputs and running livestock on bare soil. In some cases, the pressure becomes unmanageable; in the worst cases, they finally crack and take their own lives.

Driving though drought-ravaged areas in Eastern Australia in the Summer of 2019, it struck me how farmers are willing to suffer and blame external forces, rather than ask a regenerative neighbour, "How are you still growing grass and fattening cows?" Everywhere I travelled, it rained. Water came sheeting off the land, and on the radio the announcers were celebrating that dams were finally filling up. The regenerators I visited, had empty dams, as water soaked back into their parched soil. Where it belongs.

Mindset offers one of the biggest opportunities and biggest limitations; mindset informs your beliefs, thoughts and actions. "If you think you can or think you can't, you're right!" One of the most powerful speakers I've heard on this topic is Doug Avery, author of The Resilient Farmer. In his broad New Zealand twang, Doug speaks in a direct and humorous manner. He believes that the most important part of the farm to invest in is the "top paddock," the one between your ears. Doug learned the hard way, about the need to tend to the health of his top paddock. Thinking he could 'tough it out' through major depression, his farm went through 8 years of drought and financial struggles. By 1998, both the farm and Doug were on their knees. In 2013, his breaking point came while clearing trees that had been thrown by hurricane force winds earlier that month. As he sat in his excavator, the site became ground zero for a magnitude 6.6 earthquake. When the quake struck, his instant thought was that he was having a stroke, as the controls in the grapple bucket began to shudder violently. Looking across his fields, he saw sheep being tossed from side to side and he experienced a short-lived moment of relief that it wasn't just him shaking. He returned to the house and the final straw broke, half of their house had turned into a pile of rubble.

Doug, with some prompting from his wife, realised he could no longer continue to bottle things up. He needed to speak to someone and share what he had been going through. That moment changed his life for the better and he hasn't looked back since. Between his book and his entertaining speaking engagements, Doug is having a major impact in shifting the conversation around resilience and depression. "What is your current story? What's your old story?" he asks the audience. "The only real story that matters is: What is your next story?" These are the things we have control over and a key to being Regenerative is the focus on not being bound by old stories and old expectations. Doug was told he could bounce back from the earthquake from a friend, to which he replied, "Nah, I'm going to bounce forward!" With bouncing forward, he believes "you'll either win or learn." He believes we all need to take time out every day to reflect or just sit quietly, away from the busy-ness of motion, to distract us from what's really important.

Rural communities are in trouble. The average age of farmers in developed countries sits between 55 and 58 years of age; modern agriculture is failing to renew itself. Fostering and mentoring the next generation so they don't bolt to the city, is an essential component in regenerating farms and ranches. Across the planet, we are seeing a rise in global awareness around food production, animal treatment and environmental concerns. While there is a backlash against the industrial food model, regenerative agriculture provides a source of inspiration and deep fulfillment. The next generation coming through, often feel let down by their elders and are the generation with the most who are disconnected from land and food in the entire history of humankind.

Meeting young people, like 12-year-old Montanan Maloi Lannan, who are inoculating others with the soil bug, make my heart sing. Maloi has created a kids colouring book titled, "Don't call it dirt!" Motivated by an experience with an elderly vet, who was judging her school project about soil and animal health. He told her; "soil health has nothing to do with animal health," and marked her project down. This is one girl you don't want to mess with; she's determined, articulate and confident. She immediately contacted Gabe Brown, Joel Salatin and most kindly, she included me, in her quest for more information and science. Her colouring book is a delight and she takes it into schools to "teach the young kids about soils and regenerative agriculture." She plays an active role in the animal management on their family farm, Barney Creek and regularly posts updates of her learning on social media. The sales of her book enabled her to start her own flock of 10 sheep. She is providing inspiration to a wide community.

At 19, Kate Indreland was my travel companion for a 4-month tour visiting regenerative producers in North America, Australia and New Zealand. Just like their cattle, the Indrelands raised level-headed kids anyone would be proud to have. And like their cattle breeding programme, their daughter Kate is petite and tough too. She's currently studying psychology at Montana State University under the ROTC (Reserve Officers' Training Corps) programme. Today she rang from campus; she'd been up since 5am training. Training which involved lugging a 50-pound rucksack for 2 miles, before rowing a raft upstream. Her group, which was chosen from the smallest in the group, had "whipped" the other teams made up of 6-foot monstrous framed men. As a reward for their victory, their team inflicted a penalty on the other squads, pushups and squats, in the near freezing Montana waters. In her style, which always looks out for the underdog, Kates team also participated in the penalty. Imagining myself in this scenario, she lost me at the 5am wakeup call!

Travelling together through the US in my horse trailer, provided fodder for future stories. Staying in creepy fairgrounds, we had restless nights, awakened by squealing car tires and police siren, crossing our fingers that the horses would still be in their stalls in the morning. Heading to friends on a Nez Perce reservation, I overheated the truck on the White Bird Hill in Idaho. Now in my defence the White Bird Summit road climbs 825 m (2,700 feet) in over 11 km (7 miles), and there are NO warning signs, as you leave the long peaceful river valley and hit a 7% grade road. No warnings at all.

We would roll up to my friends and clients, unload horses and then I'd revel in her glee, as ranchers enveloped her open-spirited nature. The experiences from staying with producers, like the Haggertys and the Osbornes in NZ, were priceless. Her take-home message from our adventures, wasn't just a deepened understanding of soil processes. For her, it was "seeing how families dealing with crisis can still be happy. "The best part," she says, "was talking about creation and possibilities, not complaining about the weather or the neighbours. Instead, we talked about what we could do for the future." It occurred to Kate, that these people were "totally selfless." She came back with a realization that, "here is my generation's opportunity, to be the new pioneers of the west."

In 2018, she was a finalist for the General Mills "Feeding Better Futures" scholar programme, winning $10,000.[139] She's so talented and passionate about so many things, pinning down her next step is always challenging. A roper, a military liaison? An agronomist or a rancher. That is the *one* downside to a passionate mind! Last year she was set to head to Florida to train as a pilot, to top-dress biological stimulants and bring rangeland back to life. What inspires her to come back to agriculture, is "seeing families happy, seeing that their kids want to come back" with a new awareness that "profitability is possible, with far less stress on families and getting back to what people really care about." A breakthrough Australian study by the Vanguard Business Services Group into the overall wellbeing of regenerative producers, indicated that they were more profitable and "experience a meaningful and significant wellbeing advantage" compared to similar graziers.[140] With whatever Kate chooses to do in life, it promises to be something impactful. She is part of a renewed and vibrant energy which regenerative agriculture is returning to land.

Matthew, Di and Ian Haggerty's 16-year-old son has grown-up surrounded by regenerative ag and he sees himself returning to the farm when he finishes studying. His parents behaviour is one for bemusement: "They get excited about the weirdest things. The other day Mum came excitedly running into the kitchen to show me some fungi growing on manure!" He believes that as more people become aware of what's in their food, then this social trend will drive farm practices to produce healthier, cleaner, nutrient dense foods.

Roger believes the success they have had in engaging with youth and their community, comes down to the fact that their local community is increasingly open-minded and "we have no mission to think we are right." The couple have zero ego attached to outcomes or pushing their viewpoint on others and "ultimately people appreciate that." They run regular open days and host different organisations for events, attracting a diverse group of people and businesses. Their open-door policy means every lunchbreak offers a different conversation. Today, I could be sitting beside a 17-year-old Australian jillaroo[25], a hunter from the city, the local vet, a young cropping guy, or a sun-beaten cowboy with a flamboyant moustache. The Indreland's

---

[25] Jillaroo- a young woman from a cattle station.

kind and generous spirit, means all are welcome to discuss any topic without reservation or judgement.

The Indrelands have a board of advisors, which include their awesome (and humble) soil agroecologist, a ranch business trainer, renowned cattle expert and a vet. The board meets twice a year and have permission to ask anything and push the Indrelands comfort zone. As a result, profitability, enjoyment, pasture quality and animal performance are all on the rise—factors the local community is interested in learning more about.

The regenerators I've witnessed having a significant impact in shifting 'social norms,' are those who are unprejudiced and are able to see other points of view. Jono Frew, the passionate cropper based in Canterbury New Zealand, believes a shift to regenerative practices is the only option for a region that has been suffering under crippling droughts, water restrictions and nutrient legislation. He is frank, he does not believe that a crisis is enough to influence change. "There have already been plenty of crises. How big a crisis do we need?!" For him, the catalysts for a shift in practices have been community driven and are happening quickly. With a long family history in the region and wearing his rugby 'stubbies,' (the short-shorts which are staple uniform for New Zealand farmers), he commands a level of trust in conversations around the consequences of chemical use. He is unabashed about having conversations and getting interested in what other farmers are dealing with. His open nature has sparked an interest in those around him and he now finds farmers will approach him at the pub to ask questions. He's also able to shrug off any concern from critics; he's seen the results for himself and is committed to supporting others in achieving their goals. Jono is an exuberant, self-expressed personality whose curiosity and joy about regenerative agriculture, can be quite contagious.

He was not always so excited about agriculture, however. Born third generation into a chemical application family business, he tells of finding farming becoming increasingly "dull and boring", with its plug and play prescriptions. A series of 'aha' moments altered his world. The first epiphany came through in the love for his local river that he had swam and fished in as a child. Over a five-year period, the flow reduced, water quality declined, algae bloomed and fish disappeared. Today this river has dried up. At the time, Jono was the manager for a dairy farm that was drawing water from the river. Every time he turned on the pump, he says, "I began to feel part of my soul was being sucked away."

While working for his father, Jono would apply many different chemicals, including the organophosphate lorsban (chlorpyrifos). When unblocking the spray unit, Jono would begin to feel nauseous. He never wore gloves, although his Dad did provide overalls so they would look professional. When Jono asked about chemical protection, his Dad explained "Son, we don't need gloves; that stuff only kills insects."

Many farming cultures continue to believe you're weak or less than a man, if you don't tough things out. Chlorpyrifos was first developed as a nerve gas in WW2. It has been linked to autoimmune and lung diseases in adults. For children living in rural areas or exposed to chlorpyrifos, they can be impacted by a range of neurological problems leading to poor motor function, ADD and lower IQ. Use of chlorpyrifos in households, was banned in the U.S. in 2001, 2008 in Europe, 2010 in South Africa and is totally banned in Sweden. In NZ, however, it remains under government review, as a listed priority chemical. Globally in agriculture, chlorpyrifos remains one of the most "widely used organophosphate insecticides."[141]

Jono began to question chemical practices more when his grandfather unexpectedly passed away and his father, who has never smoked, is plagued by a persistent cough. He has always felt a sense of discomfort around the inputs being pushed by the industry. Attending his first regenerative soil workshop, he found the experience "incredibly confronting," as pieces of the puzzle began to fall into place. He's also the first to admit he has always been proud and stubborn, and it took some courageous reflection to realise that he'd been part of the problem he was seeing in his community.

The impact Jono and his close circle of regenerators is having on their community, is truly inspiring. He is seeing a shift in family dynamics, as "kids want to spend more time with us." He is now working with the seed producer Simon Osborne and "some days the farm feels like we're three-year-old boys running around full of excitement!" Part of this excitement, too, is seeing that his community is becoming more receptive to learn more— despite being warned by a chemical salesman to "settle down!" with his message.

Instead of being limited by community mindsets, Jono has been part of a regenerative Ag network called 'Quorum Sense.' A group that has been hugely successful in shifting social norms and encouraging the local community to engage in deeper conversations and actions. The achievements of this group and others, is in part why I don't buy into the bleak outlook that a crisis is needed. Many of the stories I have shared with you, are from people who were not motivated by crisis; they were motivated by an epiphany, or from seeing an opportunity. They have a sense that the tsunami is coming, and it is those who are willing to adapt and innovate who will be the ones riding the wave.

If we want to turn this ship around, we need to return to nature's basic rules: diversity, groundcover and sunlight capture. And we need to stop killing the most essential livestock, our soil and gut microbes, to restore integrity back to a failing system. Rebuilding the soil microbial bridge is central to producing foods with the full spectrum of vitamins, minerals and enzymes. The quality and the diversity of what we are eating is driving the insidious slide in human health. The wider human population may not be interested or aware of the declines in food quality. However, water issues, climatic variability and wildfires are now impacting the lives of so many. With a younger generation feeling increasingly disillusioned and disempowered, choosing regenerative food and fiber may just tip the balance towards transformative change. It is the consumer who can help to drive the changes; consumers like you.

We humans are a complex bunch. I am privileged to spend most of my time lodging with ranching and farming families. Staying in the homes of people from diverse walks of life, is a privilege and a reminder to always leave my assumptions at the door. It's helped me to deepen my sense of what's important to people. What do you care enough about, that would make soil health worth investing your valuable time, energy and money on?

I've found that if people aren't connected emotionally to the outcome, they won't shift the habits of a lifetime. It's often not the doing, but the starting that terrifies most. You can show people graphs and financial data about changing a practice, but it makes little impact, until they connect with their own drivers. And money isn't generally the driver. People often seek money as an end to some other means, like making a difference with their family, leaving a legacy with their land, taking that holiday, impressing the neighbours, feeling successful or feeling happy. (News flash: money doesn't make you happy—unless you're below the poverty line.)

I once met a cropper who told me his motivation to change was to be the 'Gabe Brown' of the South. I was thinking he meant planting cover crops, diversity and the building of soil, but no ... what he meant was that he wanted to be famous! Having lofty goals like this, can be real motivating factors, although I'm pretty sure if you look at most soil advocates around the world, they didn't get there to be famous. But who am I to judge?!

What provides individual motivations and catalysts to change, is as varied as the landscapes we inhabit. Producers in this book have experienced different breakdowns and encountered fear which led to their 'aha' moments. For orchardist Nick Pattison, his motivation came from seeing organophosphate insecticides settle around his dying mother; at Cottonwood Ranch, it was the reality of seeing the ranch foreclose; and for entomologist Jonathan Lundgren, it was the realisation that agencies were attempting to silence his work on neonicotinoids. For many, digging a hole and seeing compacted or water-repellent soils, provides all the spark they need to transform how they interact with landscapes. We all have different motivating factors; mine is to connect deeply with land and people and to feel like my life has a purpose. The Indrelands are driven by producing fit-for-purpose animals, to do a good job by their buyers and to enjoy their lifestyle. The Stahl's are motivated to leave the land in better shape and be the best at what they do (and to "out Gabe, Gabe!"). What is your motivation? Everything comes back to soil. What would make soil health a worthwhile investment of your time and energy?

Trying to change is hard, particularly those deeply ingrained habits. Even when people know the costs of continuing a behaviour and even when they really care about the outcome, inertia can be significant. Consider friends, you know who want to stop smoking, lose a few pounds, eat healthier or spend more time with the kids. There is a lot of research that shows the energy required to modify a habit can initially feel greater than the benefit.

Achieving new goals requires a catalyst, or 'activation energy,' in a process like the one involved in chemical reactions. Initially, more energy is needed to take new actions, then the system resets at a lower maintenance energy. For many, it can feel like it's just not worth devoting the energy to shift current habits. We have a client who gets a headache every time we discuss new potential actions. He tells himself he's too old and not smart enough to take on new information (he's 48 years old and runs a massive, diverse farm operation!). It's this activation energy working against him, the changes all seem too overwhelming.

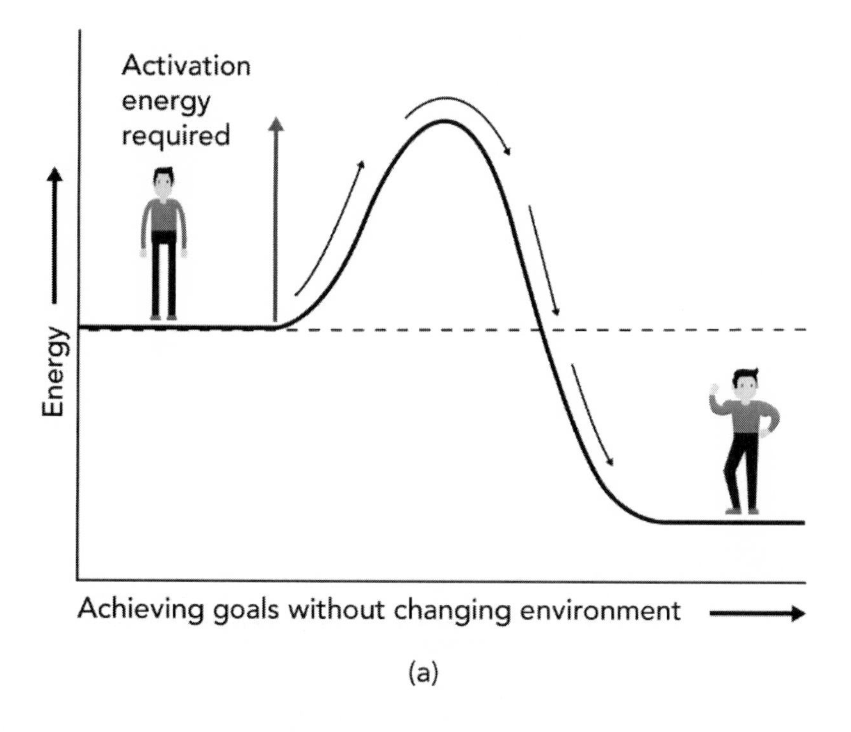

(a)

One way to reduce inertia and transform our thoughts and actions is by shifting our surrounding environment. Through doing this, new actions require less motivation and willpower (activation energy). If you want to lose weight or build muscles, don't spend as much time with pizza-eating, TV-watching friends. If you want to reduce your chemical inputs, fire your chemical agronomist, yes, the one who gives you a hat every year and receives a commission on your sales. For many transitioning into regenerative practices, this is a challenging but necessary first step. Unless your advisors are open-minded, engaged in learning and exploring novel possibilities, they will increase the activation energy you need to make changes, resulting in more headaches.

When taking the first steps, find a coach or mentor, someone in your community you can chew the fat with. The adage "when walking through the minefield, follow in the footsteps of those in front of you" holds up whenever taking steps that can feel risky.

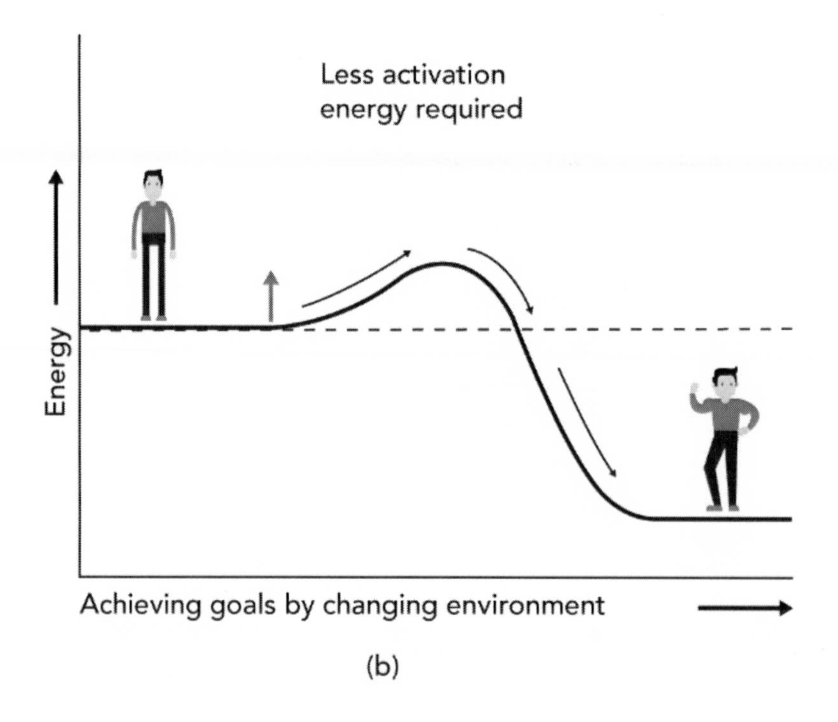

(b)

With technology making the world as small as it is now, there is no reason you can't contact producers in other countries. Many progressive producers are easy to find on the internet and a lion's share of them love to talk about their favourite topic. At my company, Integrity Soils, our business mission statement is to create clients who no longer need us. It's a terrible business model, however, it is one defining aspect of regenerative agriculture: reducing the need for external props and support. Our services shift away from recommendations, towards a focus on refining systems and providing a soundboard for new ideas. Often the biggest step required, is the activation energy to get started.

The life blood for successful regenerative practices, is ongoing learning and education. Sadly, at this stage, there are only a handful of places in the world to study regenerative agriculture. For young people eager to learn more about these practices, they flock to high-profile regenerative operations to intern or volunteer. These farm enterprises are receiving hundreds of applications from young people vying for only a limited number of places.

For myself, my biggest learning experiences arose from what we came to call 'biological barbecues.' In 2004, a group of farmers and growers would meet every six weeks at a different farm, digging holes and discussing what was and wasn't, working. We jokingly referred to it as; "ruffling through the underpants drawer," as nothing was hidden! The learnings from a group of people willing to be transparent, vulnerable, honest and engaged, provided the formative learning days of my life. I learnt more with these farmers, than I ever did from years of formal education.

The Victoria No-Till organisation in Australia, has long recognised the value of soil health, committed to education and building farm resilience. Every 4 months, they produce a fantastic glossy magazine, profiling real-world case studies from farmers in the area. Historically, the group focused on technological fixes and using controlled traffic to build soil. In my view, technology offers real, but only incremental change and benefits. Combining modern equipment with soil health principles offers a total transformation for what's possible, for soil, landscapes, food and communities. For these large-scale operators, regenerative agriculture is a 'game changer.' Chairman Grant Sims, is seeing a "real buzz in Victoria; there's so much positivity right now, which is great." As a collective, these farmers are trialing different practices, including cover crops, inputs and livestock management. These stories of success and failure are then shared in their magazine and at their annual conference. Reading their stories, their 'mastery', 'autonomy' and 'purpose,' comes shining through.

Our role as regenerators today, is to step outside of the box we've built around ourselves; from a concern of failure or "who am I to speak?" or "what will others think". In most cases, they have the same concerns as you. It is now time to share and celebrate, what has been going on underground.

Shifting our stories around 'failure', can provide powerful new insights for action and learning. It's not unusual to hear regenerators talk about failure as being one of the best experiences in learning. We've become so fearful, that the imagined risk from trying something new and different, can literally paralyse us. Fear is the body's way of keeping us physically and mentally safe; however, in agriculture, it creates the opposite outcome, decreasing the resilience of businesses. This may be why so many progressive regenerators are coming to agriculture as a second career, not limited by historic stories, fears or, "this is how we've always done it."

Travelling in my youth, I was fearless and often put myself into vulnerable, unsafe situations, that ended badly. However painful, this learning was incredibly powerful, setting me onto new paths in life. Paths that provided unexpected gifts. I've since learnt to avoid dangerous situations and I've come to understand failure differently. Today, I even get excited about breakdowns. Now, that's not to say I enjoy them, but I've come to appreciate that on the other side of every failure, or breakdown, lies a breakthrough. After the tsunami struck in Thailand, a Moken fisherman said the event, "was an opportunity; it has made a strong group, stronger."

This for me, is the very definition of Regenerative Agriculture. How do you face a challenge or view adversity? Do you see indicators in the landscape and opportunities to address root causes, or are you merely reacting?

The challenges facing the planet, do not only require food producers to do things that are different or new. These are the words of change and change still requires the same kind of thinking. Regenerative agriculture involves a transformation, in the way we see and interact with land, plants, animals and people. What we are witnessing with the regenerators, is the emergence of the butterfly, with a focus on restoring integrity, water and carbon cycles to produce nourishing nutrient-dense food. These people are also bringing vibrant life back into their communities. There is no need to be reactive, or to battle against the technological, engineered solutions, with their glut of 'unintended consequences.' Short-term, short-cut thinking, just doesn't belong in a healthy, resilient, autonomous future.

I picture regenerative agriculture functioning like a mycelial network, transforming scarcity and the blinkered and competitive mindsets to that of connected, diverse and collaborative communities. Systems where the whole is greater than the sum. There is no great leader or 'guru' with all the answers; the answers lie in the collective. Our wisdom and experiences span across different ecotypes, crops, livestock and wildlife. Everything in this book has been gifted to me from others, my community, clients, attending workshops and from past generations who put their learning onto paper. Soil health is not a new concept; we just forgot about it in the dizziness of technology and falsely believing we were smarter than nature.

Real success in land stewardship lies with biomimicry. How does nature work? Observing patterns, where pests and weed populations are shrinking, or where animals choose to graze, can provide a catalyst for new action or even to stop actions. As rancher Steve Charter once eloquently shared, "We need to humble ourselves to nature. To understand more deeply how it works. It's a beautiful and sophisticated process. Once we co-operate and work with nature, then things really start working." The 'modern' agricultural paradigm has encouraged us to stop listening to our intuition. Most producers I meet can no longer stomach the application of chemicals to food, or to witness animals being treated inhumanely. The heart of these observations rests on how we connect to each other and our landscapes. Taking the time to turn off our over-analytical minds and begin to see the world around us newly. 'Being,' instead of 'Doing.' How often do we take the time to just sit with our land, plants and animals, without being in action?

The greenwashing of the terms "regenerative" and "nutrient density" is inevitable. It has already begun. Corporates and Agri-Chem companies are seizing upon the marketing opportunities and are offering their watered-down version to the masses. Without measurement, these phrases lack integrity. There is enough wool pulling over consumers' eyes, without adding to more labeling confusion.

For those looking to market foods as 'Regenerative,' my challenge to you is to put your money where your mouth is. Test your meat, fruit and vegetables for multi-residues and nutrient density. There needs to be far more integrity across the entire food system. Evidence that you are regenerating your soil, water cycles and plant cover is essential. This integrity needs to transfer across all aspects of the enterprise. Picture that you have an urban auditor at your shoulder. Could you justify your actions to them through your working day? Many of the stewards in the case-studies I've shared, have open doors and are transparent about what they're doing on their land. They are benchmarking and monitoring changes over time and if they're marketing their products as nutrient dense, they measure it. Ensuring integrity, comes down to individuals and in our ability to share our stories honestly.

Talking to producers who promote their foods as "nutrient dense" at farmers markets, or online forums, I feel a little like the fun police. "Have you been measuring nutrient density?" I ask. "Well," they'll sheepishly reply, "it's grass fed, so it's automatically nutrient dense." How about "it's natural" or "it's free range" or "spray free so ..." Without some measurement, using these labels just becomes an empty marketing ploy. The certified, 'shallow' organic foods can leave consumers with a bad taste in the mouth and test lower in nutrients. This sadly undermines the entire organic sector.

Organic production has similar roots to the regenerative movement. Producers were looking for deeper ways to work with landscapes, improve soil and produce quality food. From there two paths emerged, the deep organic path, which sits in the regenerative space and the shallow organic. Shallow organic, is the 'light green' that the orchardist Nick Pattison pointed to in his journey. It is an approach that substitutes hard chemicals with softer, organic versions, replacing herbicides with tillage and ignoring the underground livestock. This approach has tick boxes to assure: "I don't use synthetic chemicals." However, it leaves consumers with no assurance of environmental health or food quality. The way to distinguish between the two, is by questioning producers about their management. Are they measuring soil improvements? If they are claiming their products are nutrient dense, ask to see their tests. Additional clues that a claim may not have substance come through in flavour, insect damage and the storability of fruit and vegetables. If your food has insect damage, is it because this fruit is organic, or is it because it is not fit for human consumption?

The regenerators live and breathe in a world of wonder and possibility, around different aspects of their lives. They are asking themselves, "am I being regenerative towards my animals, my business, my land, family and myself?" It's not uncommon to find that people are strong in one area and weaker in another.

Just as there is no such thing as a perfect soil, there is no perfect human being. Often the most vocal and philosophically minded, are the people who may not be as strong in the practical realm. Some who may be great livestock people, may struggle with personal relationships. I see producers exchanging their inputs from chemical to biological and not increasing their profits. Is this regenerative? This doesn't make these different approaches wrong; rather, it provides a call to action to strengthen where we're not so strong or to invite in others to support our regenerative goals. To heed the call of the mycelial network, you must acknowledge that no system stands in isolation, including you.

Take the example of Farmer B; they're fulfilling on producing a profitable, ecological, nutrient-dense product, at the cost of their human relationships and their own wellbeing. This is an insightful exercise to do with yourself and your family.

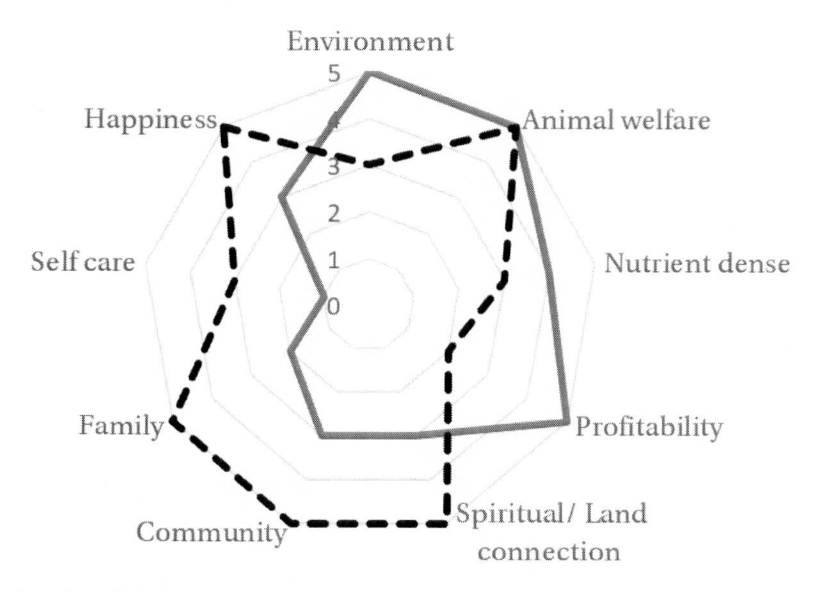

Graph: Pillars for Regenerating Landscapes

For most of us, it seems self-care takes a backseat, but who will manage your land for you if you're not there?! It's like peeling back the layers of an onion, to discover what's next. Yet another layer of learning, growth and expansion on what's possible. Regenerative agriculture is an exciting, complex and at times, challenging journey, with no end destination. Enjoy!

The wellbeing of our agricultural lands and own health are intimately linked. Every day it becomes clearer to me, that our outer environment is but a reflection of our inner world. We cannot sustain 'business as usual' in an arms race against nature and ourselves. Regenerative land stewardship offers an immediate strategy to reconnect to both our inner and outer landscapes, for those who have the will to listen more deeply.

Living in this present moment, I feel a palpable sense of excitement and joy from the conversations I hear in rural communities, ones that once seemed so bound against change. There is a global collective of deep experience and wisdom around regenerating landscapes. We don't require more technological fixes or genetic tinkering to address the challenges we face, instead we require an evolution in natural intelligence, reconnecting deeply with nature. There's no turning back once you've caught soil health bug and its' boundless potential. The invitation from nature is always there: go forth and inoculate. Pollinate your communities with the soil bug!

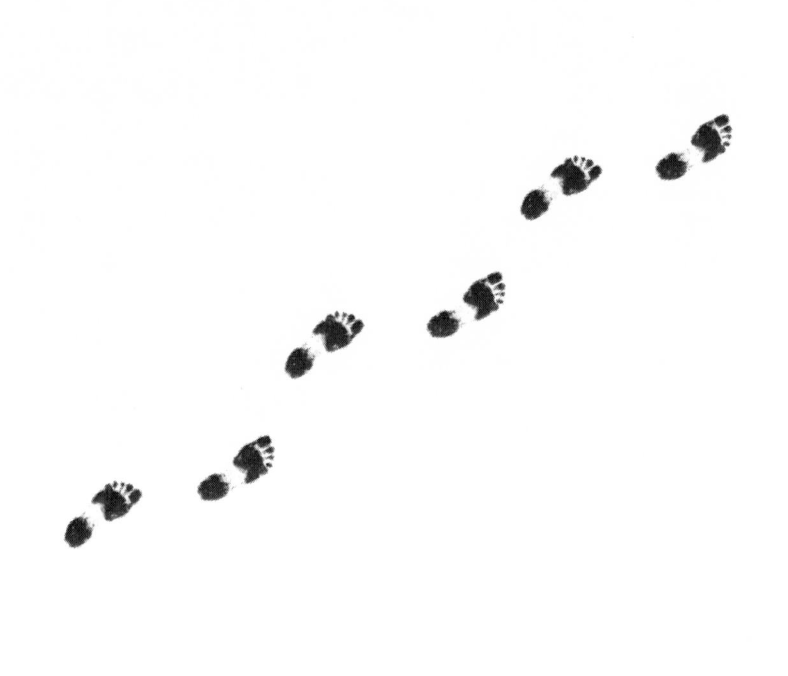

Appendix

Transitions

The drivers for change for many producers in this book were inspired by an epiphany. However, you could wait your whole life for one of those. If the actions of the regenerators are new or daunting to you, then here are some potential steps to build confidence.

1. Start with something ridiculously small.
If your concerns about risk have you sitting on the fence, set yourself up for success. Plant just a few trees. If you plan to spray for weeds, have fulvic acid ordered and sitting next to the herbicide. Try putting seeds in with your livestock mineral. Alternate a fungicide with a nutritional spray. Trial one pasture with heavier animal density and then miss it from your next rotation. Swap chemical seed treatments with a biological seed treatment. Add humic/humates to your nitrogen and drop by 30% in an area.

2. Shift your environment.

When it comes to transforming our thoughts and actions, our catalyst is the environment. We can shift the environment around us in such a way, that taking action requires less motivation and willpower (activation energy). Surround yourself with regenerative producers and create a discussion group, "biological barbeque." Read magazines and books.

3. Regenerate your thinking processes.

It is the stories that you believe or tell yourself, that give you actions or prevent you from learning. The only way to really shift our actions, starts with who we see ourselves as 'being', or our mindset about who we are. This 'being' then informs our thoughts, our language and our habits. It's important to focus on the right steps, versus the right result. Completing personal and professional development programmes, are critical steps to reveal potential "blind spots" for action. I'm a big advocate for Ranching or Grazing for Profit Schools in

North America and Australia. And if you really want to get to the bottom of your own life story, Landmark Worldwide offers training across the planet.

4. Get out of nature's way!

The key to regenerative agriculture is not the answers, but the quality of the questions we ask. One excellent habit to develop, is to ask the 'why' question at least 6 times, like; "Why are you calving in the mid of Winter? Why are you cutting hay? Why are you using that insecticide, or growing a crop in an unfavourable environment?" I am amazed at how often the last answer becomes "because that's how we've always done it." Asking these why questions and then taking action, can be the most profitable exercise you could ever undertake.

Ask the question; "How would this work in nature?" For instance, "When are the wildlife having their babies? How does this tree or vine grow in the wild? How can we align our practices with nature's basic rules?" The further we stray, the bigger the costs. Some of the most powerful actions we can do to regenerate land, are to stop doing the things that are causing harm or working against the sync of nature.

5. Have patience dear friend.

This may be the most challenging aspect, for those accustomed to the rapid hit from chemical inputs. How long has it taken to degrade your land resource? The changes in the soil, may happen long before benefits may be visible above ground. Humans by nature are impatient, but if we dig up a seedling to see if it has germinated, we undermine its progress. Just trust the seed is growing and visible changes are coming. Monitoring soil health, infiltration, plant tissue tests and working alongside a mentor, helps to alleviate your concerns. There is a certain degree of faith and trust applied in the transition. Understanding the "unintended consequences" of the quick fix and the breakdown that is occurring at every level of our current modern farming model, can help steady our focus. The science is there, consumers are waiting and the political awareness and will for action is now bubbling up.

Transition steps

Avoid costly production losses by building on local knowledge: Mentor; consultant, a successful farmer, or join a discussion group. Do your own trials.

Education: Books, courses, workshops, podcasts...

Benchmark: Measure where you are now; soil mineral, biology, carbon, leaf tests and photos. **Hone your observations.**

First do no harm: Reduce and then eliminate products that blow the microbial bridge; soluble N and P, herbicides, fungicides. Buffer chemicals with microbial foods.

Triage and address major limitations: Sunlight capture, air, water, decomposition and minerals.

Encourage biodiversity above and below ground: Herbal leys, fodder crops, cover crops, shelter belts, inter-planting. **diversity, diversity, diversity!**

Implement practices: Increase photosynthesis, rooting depths and soil carbon. **Avoid bare soil at all times**.

Apply broad-spectrum products: Feed biology and address major nutrient deficiencies, i.e. Lime, rock phosphate, guano, seaweed, fish, seawater, compost, vermicast, sugar etc.

Health: Ensure crop and animal health needs are met; if not, use free choice minerals, probiotics etc. Test quality of your produce.

Monitor and observe changes: Brix, EC, pH, photographs. Adjust programme if required.

Strategic long-term planning: Land and stock use choices. Focus on business resilience.

When you can see your successes, **mentor others**!

These tips were created from brainstorming sessions with over 300 regenerative farmers, growers, researchers, educators and support companies during the Association of Biological Farmers National Roadshow with Dr Christine Jones, June 2010.

Using a refractometer

A refractometer measures degrees, or %, Brix – the dissolved solids and the sugars produced during photosynthesis.

It takes less than a minute to give you a reading of the status of your crops. Measuring plant sap Brix across several samples in a crop, gives an immediate insight into the general health of that crop. Take multiple readings consistently, at the same time of the day, from the same part of the plant (then ensure all subsequent readings follow suit). Record your findings and compare the trends! Start a habit of recording as often as possible, while you build a picture of your land.

Brix levels will change through the day, starting lower in the morning, peaking early to mid-afternoon, before dropping with the sun. This is consistent with the daily plant growth cycle and the absence or presence of sunlight. Generally, optimal health and quality for grasses, is indicated by a reading above 12° and for most legumes, above 14°.

Brix can help determine the suitability of a foliar spray.

Methodology:

Measure Brix on a crop, before a foliar spray application and on a control area with no application to be applied. Re-test both areas, 1-24 hours after application of different foliar sprays.

Brix levels needs to lift by at least 2 points above the control to be considered suitable for application. If it remains the same or drops, then re-test the Brix level in one week. If the Brix level is still low after a week, then it's reasonable to conclude, that the application of these inputs, are not effective at this time. Use the 5M approach to reconsider what may be the limiting factors. Mindset, Management, Microbes, Minerals and OM.

Operating Instructions

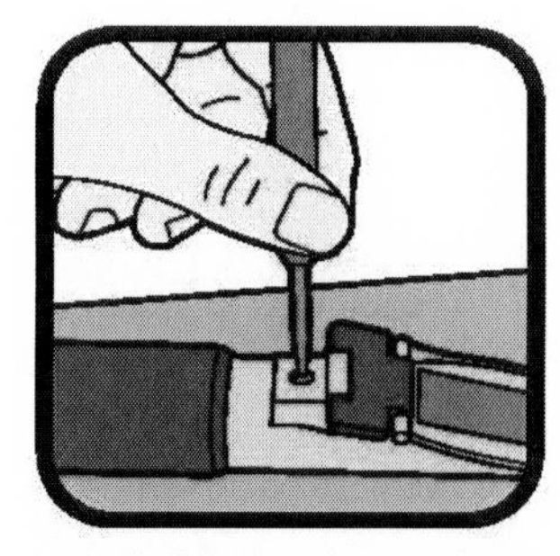

Calibrate refractometer using pure, distilled water

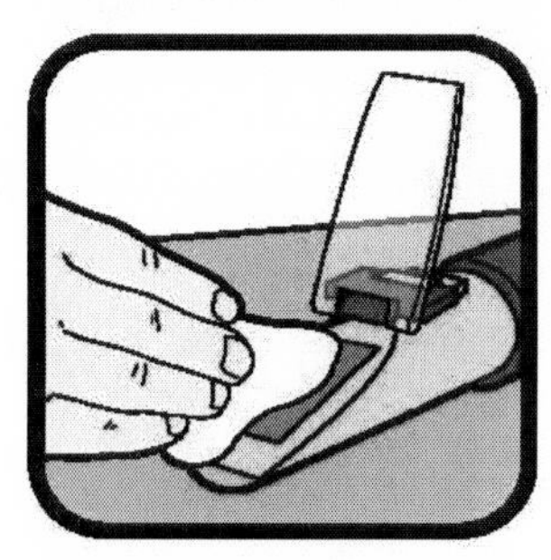

Open the daylight plate, wipe the refraction prism carefully with soft flannelette. Be careful not to scratch the surface.

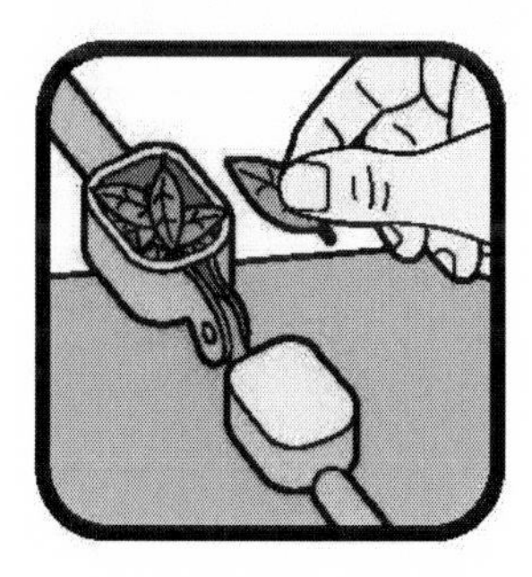

Choosing the fully developed leaves, twist the leaves a few times and then place into the well of a good quality garlic crusher (Jumbo Zyliss) (if sampling low sap species like avocados and grapes use sap vice grips).

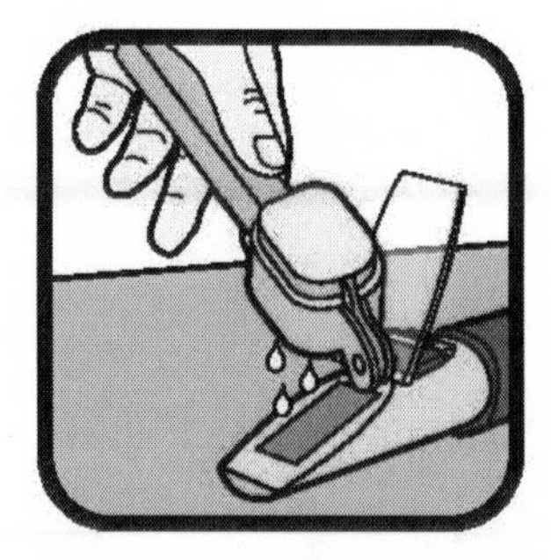

Squeezing out the sap. (If plant material squeezes out of the crusher –place a coin in the bottom of the well). Put 2-3 drops of sap on the prism surface

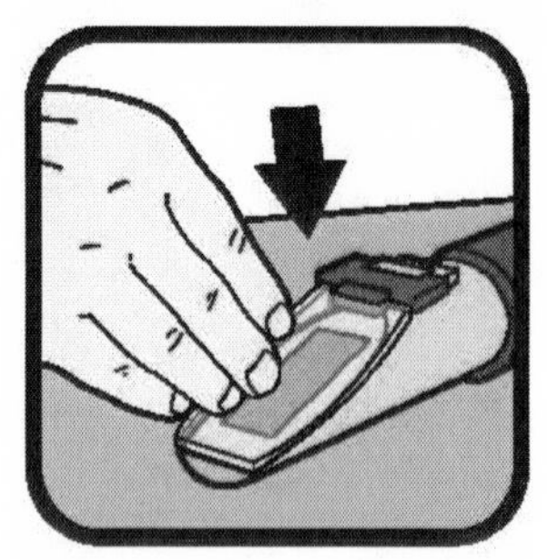

Cover the daylight plate slowly to let the solution cover the whole prism surface reducing any air bubbles.

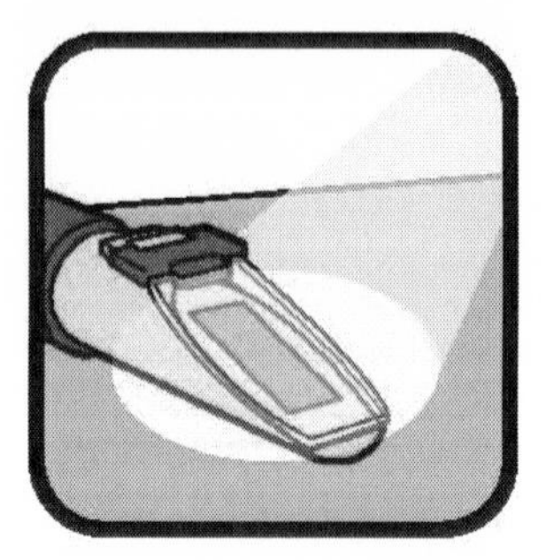

Turn the refractometer towards a light source or bright place.

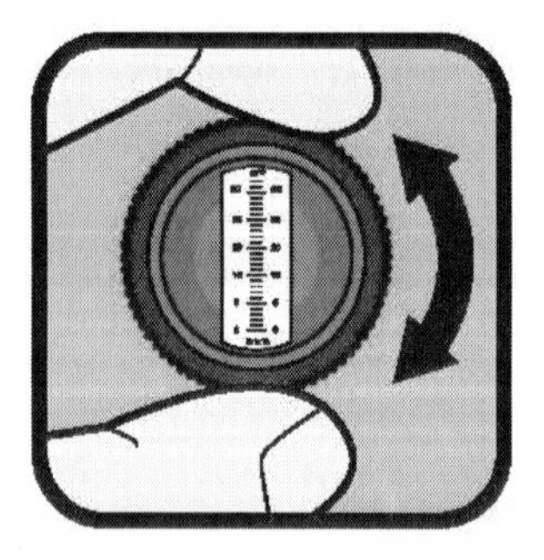

Look through the lens.
Turn the focus adjustment until the graduated lines can be seen clearly.

MORE ON BRIX

Brix levels can vary due to stress and/or dehydration and once plants set flower/seed. Therefore, it is vital to keep good records and monitor changes in Brix level over time and to avoid sampling insect or disease damaged leaves.

The method of extracting sap can have a large influence on the reading.

Generally, Brix readings will drop with low atmospheric pressure e.g. the onset of a storm.

Analyse your weeds! The sap of weeds should have a lower Brix level than the crop. If this is not the case, you need to look at why your current management is favouring weed production. If Brix is higher in your weeds, you may need to intervene to reduce the threat of yield reductions. If the Brix level is lower in the weeds, you do not need to step in, as the crop will out-compete them in time.

The lower the humus levels in soil, the faster the Brix level will drop following prolonged cloudy or rainy periods. In these conditions, to prevent Brix level dropping quickly use a foliar spray of fulvic acid.

Brix levels should be uniform when sampled throughout the plant; if not, then suspect a soil imbalance. P:K ratio is a key suspect here.

Brix levels track the intensity of the sun, so are lower in the morning than afternoon. If Brix remains stable through the day, then suspect boron deficiencies. This critical trace mineral is responsible for translocating sugars between roots and leaves.

As with all testing, Brix levels need to be weighed up with other tools and good observations. It is not a tool to use in isolation.

Take good records, during any single day and across the season, to get a true picture of your crops.

Infiltration Test

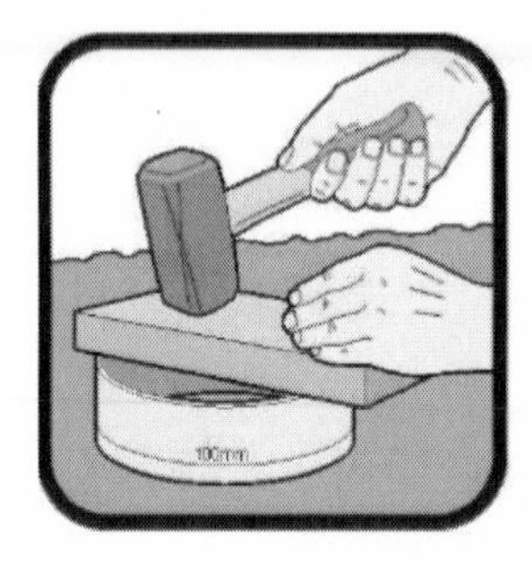

Place a piece of wood over the top of the ring.
Strike the board with the hammer until the ring is driven into the ground to the 100mm mark.
If the soil is very dry and compacted or rocky, then use the knife to cut a slit into the soil for the ring, whilst disturbing the soil as little as possible.

Line the soil surface inside the ring with a sheet of plastic wrap t completely cover the soil and ring. This procedure prevents disturbance to the soil surface when adding water.

Insert a ruler and add 25mm (1inch) water.

Slowly pull the plastic sheet away and start your TIMER. When the last bit of water disappears check the time again and see ho many seconds / minutes it took for 25mm of water to soak in.
Using the same ring, repeat Steps 2, 3 & 4 a further 2 times.
Record the time elapsed for the each infiltration measurement.
All of the tests should be conducted consecutively.

If you conduct multiple tests and they produce the same result, this result is most likely an accurate estimate of the saturated infiltration rate.

Infiltration considerations:

If the soil is saturated, the infiltration test will not work, so wait a few more days for the soil to dry out.

If you capture an additional 25mm rain, it can mean 1/2 ton/ha (500lbs/ac) in extra yield.

If the soil surface is uneven inside the ring, count the time until half of the surface is exposed and just glistening.

The moisture content of the soil will affect the rate of infiltration; therefore, two or three infiltration tests are usually performed (if soil is dry). The first inch of water wets the soil and the second inch gives a better estimate of the infiltration rate of the soil.

Even a high-quality soil, will eventually break down after several storms. The aim is to improve soil structure to maximise infiltration and minimize run-off.

Monitoring Indicators

Soil indicators	**Plant indicators**
Soil structure/ porosity	Sap Brix/EC/pH
Colour and # of mottles	Plant growth
Soil Colour/carbon	Plant diversity
Earthworms/dung beetles	Legumes (nodules red)
Soil smell/taste?	Weeds/pests/disease
Infiltration rates	Plant colour
Water holding capacity	Urine/manure patches
Surface relief	Pasture utilisation/palatability
Temperature	Rhizosheath development
Penetrometer (compaction)	Root length and root density
Soil pH, EC (electrical conductivity)	Area of bare ground
Soil mineral tests	Drought stress
Soil biological testing	Input costs to maintain production
Enzyme activity	Plant tissue tests: minerals, RFV, ADF, Crude Protein...
Respiration	Storability and digestibility
Fungi:Bacterial ratios	Near infrared spectroscopy
Mycorrhizal colonisation	
Biological diversity	

Growing Mycorrhizae

Make your own mycorrhizal inoculant – combine potting mixes with root materials collected from plants in local healthy ecosystems. *Grow AMF spores in this potting mix using C4 grasses, such as Sudan grass, Paspalum spp, Corn etc.*

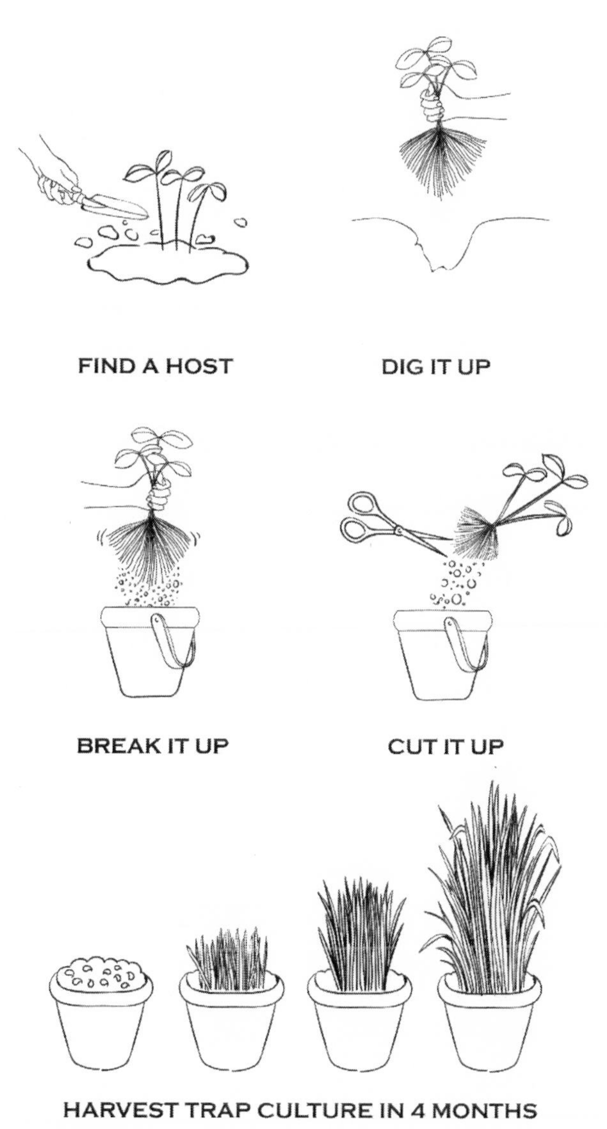

Transition weed strategies

Brewing weed teas

Fill a drum with your troublesome weed, place a heavy weight on top of the weed and slowly let them stew in their own juices. If the plant contains high cellulose or lignin, then you'll need to add some water to help it break down. Have a tap on the bottom of the drum and extract the liquid to spray onto heavy infestations. Sometimes we've seen extraordinary results and then other times nothing!

Trial different rates from a few 100mls to 20 litres/Ha (2 gal/ac). My thinking is you're breeding up the bugs that like to eat your weed and you're also concentrating the trace elements that this weed has been accumulating. (Thanks to Steve Erickson for this recipe.)

Brewing allelopathic chemicals--harvest mature plants with known suppressive chemicals, such as Forage radish, Sorghum, Sunflower or Mulberry. Dry and chaff these plants before soaking them in unchlorinated water for 24 hours. Filter the liquid and apply as you would with a herbicide. Results have shown this method can be more effective than chemical controls. You'll need a lot of plant materials for this, around 25kg/ha (lbs/ac).

Weed Indicators

Bare soils
Scrambling fumitory (*Fumaria muralis*), Purslane (*Portulaca oleracea*), Spotted spurge (*Euphorbia maculate*), Caltrop (*Tribulus terrestris*), Hedge mustard (*Sisymbrium officinale*), Field bindweed (*Convolvulus arvensis*)...
Low organic matter
Catsear (*Hypochaeris radicata*), Cape daisy (*Arctotheca calendula*), Hawkweed (*Hieracium caespitosum*), Cheatgrass (*Bromus tectorum*), Yellow alyssum (*Alyssum alyssoide*), Spotted knapweed (*Centaurea stoebe*), Leafy spurge (*Euphorbia esula*)...
Compaction or surface crusting weeds
Dock (*Rumex spp*), Canadian thistle (*Cirsium arvense*), Buttercup (*Ranunculus repens*), Wild chamomile (*Matricaria chamomilla*), Clubmoss (*Lycopodium*), Pennyroyal (*Mentha pulegium*) and Sedges and Rushes...
Nitrate and release valve weeds
Capeweed, (*Arctotheca* calendula), Black nightshade (*Solanum nigrum*), Kochia (*Kochia scoparia*), Nettles (*Urtica dioica*), Fat hen/Lambsquarters (*Chenopodium album*), Foxtail barley grass (*Hordeum jubatum*), Russian and Milk thistle (*Silybum marianum*)...
High available potassium and low phosphorus
Many broadleaf weeds: Dandelions (*Taraxacum*), Common plantain (*Plantago major*), Scotch thistle (*Onopordum acanthium*), Tansy mustard (*Descurainia pinnata*), Wild radish (*Raphanus raphanistrum*), Common purslane (*Portulaca oleracea*), Nightshade (*Solanaceae*), St. Johnswort (*Hypericum perforatum*) and Inkweed (*Phytolacca octandra*)...
Early successional bacterial species: Foxtail barley (*Hordeum jubatum*), Quack/Couch (*Elymus repens*), Wild oat (*Avena fatua*), Cheat grass (*Bromus tectorum*), Barnyard grass (*Echinochloa*),

Johnson (*Sorghum halepense*), Kikuyu (*Pennisetum clandestinum*), Medusahead Rye (*Taeniatherum caput-medusae*). **Allelopathic plants:** Water hyacinth (*Eichhornia crassipes*), Garlic mustard (*Alliaria petiolate*), Knapweed (*Centaurea stoebe*), Kochia (*Kochia scoparia*), Black walnut trees (*Juglans nigra*), Sorghum (*Sorghum bicolor*), Mulberry (*Morus*) and Ryegrass (*Lolium*)...
Fungal, or 'sleepy', soils Blackberry (*Rubus spp*), Wild rose (*Rosa spp*), Poison Oak (*Toxicodendron diversilobum*), Hemlock (*Conium maculatum*), Foxglove (*Digitalis*), Hollyhock (*Alcea rosea*), Hawkweed (*Hieracium*), Hemlock (*Conium maculatum*), Bloodroot (*Sanguinaria canadensis*), Wormwood (*Artemisia absinthium*), Mullein (*Verbascum spp*), St. Johnswort (*Hypericum perforatum*), Houndstongue (*Cynoglossum officinal*), Matagouri (*Discaria toumatou*), Bracken (*Pteridium spp*), Gorse (*Ulex*), Broom (*Cytisus scoparius*), Rabbitbrush (*Chrysothamnus spp*), Sagebrush (*Artemisia spp*), Willow (*Salix spp*), Sweet briar (*Rosa spp*), African Boxthorn (*Lycium ferocissimum*) or Mesquite (*Prosopis spp*).
Non-mycorrhizal Disturbed habitats, with low competition with other plants and high soil phosphorus; for example, *Amaranthaceae, Brassicaceae, Caryophyllaceae, Chenopodiaceae, Polygonaceae* and *Urticaceae.* *Cyperaceae* (Sedges); *Juncaceae* (Reeds); *Brassicaceae* (Fanweed, mustard, Kochia, Shepherds' purse); *Amaranthaceae* (Pigweed, Beets, Spinach, Lamb's quarters, Quinoa. Russian thistle or tumbleweed (*Salsola tragus*); *Urticaceae* (Stinging nettle); *Lupinus* (Lupins) and *Proteaceae* (Macadamia, Bottlebrush). In P-limited environments, species have evolved specialized P-mining roots without MF, including *Cyperaceae, Haemodoraceae, Proteaceae* and *Restionaceae.* As a general rule, if a plant is carnivorous, lives in water, or is a parasite in a tree, it has evolved without a mycorrhizal relationship.

Index

Bibliography

Introduction

[1] www.scientificamerican.com/article/only-60-years-of-farming-left-if-soil-degradation-continues
[2] Ong, E. K., & Glantz, S. A. (2001). Constructing "sound science" and "good epidemiology": tobacco, lawyers and public relations firms. American journal of public health, 91(11), 1749-1757.
[3] Ikigai image by @emmyzen (Emmy van Deurzen) source: https://t.co/TiRhcMD7HP

Chapter One

[1] Meng, Zhiyuan, et al. "Perinatal exposure to Bisphenol S (BPS) promotes obesity development by interfering with lipid and glucose metabolism in male mouse offspring." *Env research* 173 (2019)
[2] https://www.gq.com/story/sperm-count-zero

Chapter Two

[1] "The Sea Gypsies." 60 minutes, March 20, 2005.
[2] Taking responsibility for complexity. www.odi.org/sites/odi.org.uk/files/odi-assets/publications-opinion-files/6811.pdf
[3] www.oecd.org/environment/environmental-pressures-rising-in-new-zealand.htm
[4] www.pmcsa.org.nz/wp-content/uploads/PMCSA-Freshwater-Report.pdf
[5] Amundson, R.; Berhe, A. A.; Hopmans, J. W.; Olson, C.; Sztein, A. E.; Sparks, D. L. (2015). "Soil and human security in the 21st century". Science. 348 (6235): 1261071–1261071
[6] https://www.reuters.com/article/us-eu-morocco/western-sahara-dispute-should-invalidate-eu-morocco-fish-deal-eu-court-adviser-says
[7] "NOAA: Gulf of Mexico 'dead zone' is the largest ever measured". National Oceanic and Atmospheric Administration (NOAA).
[8] www.washingtonpost.com/science/a-major-punch-in-the-gut-midwest-rains-projected-to-create-gulf-dead-zone/2019
[9] http://www.tropentag.de/2005/proceedings/node181.html.
[10] http://www.stuff.co.nz/business/farming/4963565/Search-for-reasons-for-bucolic-beauty

Chapter Four

[11] Soudzilovskaia, N. A., van Bodegom, P. M., Moreno, C. T., van't Zelfde, M., McCallum, I., Fisher, J. B., & Tedersoo, L. (2018). Human-induced decrease of ectomycorrhizal vegetation led to loss in global soil carbon content. bioRxiv, 331884.
[12] Scarcella, A. S. D. A., Bizarria Junior, R., Bastos, R. G., & Magri, M. M. R. (2017). Temperature, pH and carbon source affect drastically indole acetic acid production of plant growth promoting yeasts. *Brazilian Journal of Chemical Engineering*, *34*(2)
Leake, J., Johnson, D., Donnelly, D., Muckle, G., Boddy, L. and Read, D. (2004) Networks of power and influence: the role of mycorrhizal mycelium in controlling plant communities and agroecosystem functioning. Canadian Journal of Botany, Volume 82, Issue 8, 2004, pages 1016-1045
Catská, V. (2018) Interrelationships between vesicular-arbuscular mycorrhiza and rhizosphere microflora in apple replant disease Biologia Plantarum, Volume 36, Number 1, 99-104.
Rillig, M. C., Ramsey, P. W., Morris, S., & Paul, E. A. (2003). Glomalin, an arbuscular-mycorrhizal fungal soil protein, responds to land-use change. Plant and Soil, 253(2)
Kazda, M., Salzer, J., Schmid, I., & Von Wrangell, P. (2004). Importance of mineral nutrition for photosynthesis and growth of Quercus petraea, Fagus sylvatica and Acer pseudoplatanus planted under Norway spruce canopy. Plant and soil, 264(1-2), 25-34.
Repka, J. (1983). The effects of mineral nutrition on the photosynthetic and respiratory activity of leaves of Winter wheat and maize varieties. In Genetic Aspects of Plant Nutrition (pp. 279-283). Springer, Dordrecht.
https://www.researchnester.com/reports/biofertilisers-market-global-demand-analysis-opportunity-outlook-2024/193

Wang, L. (2001) Fungi slay insects and feed host plants.
American Society of Agronomy (ASA), Crop Science Society of America (CSSA). "Where has all the soil gone? Focusing on soil loss important to researchers." ScienceDaily. ScienceDaily, 18 June 2014.
Bleam, W. F. (2012) Chapter 6 - Natural Organic Matter and Humic Colloids. Soil and Environmental Chemistry, Pages 209-256.
Dixon, E. F., & Hall, R. A. (2015). Noisy neighbourhoods: quorum sensing in fungal–polymicrobial infections. Cellular microbiology, 17(10), 1431-1441
Lefroy, E. 1991. Native grasses in south Western Australia. Proceedings of the Native Grass Workshop
http://www.sciencemag.org/news/2012/08/amazon-seeds-its-own-rain
https://www.nature.com/news/2008/080228/full/news.2008.632.html
Taschetto, A. S., & England, M. H. (2009). An analysis of late twentieth century trends in Australian rainfall. International Journal of Climatology, 29(6), 791-807.
http://austhrutime.com/west_australian_wheat_belt_salinity.htm
http://soilquality.org.au/factsheets/soil-acidity
http://farmdocdaily.illinois.edu/2016/09/international-benchmarks-for-wheat-production.html
http://www.abc.net.au/7.30/tragedy-underlines-wa-farming-hardships/4651942
Calculation based on 13,000 ha x 2T = 26,000T C x3.67 = 95,420 T CO_2e. Americans annual CO_2 emissions 18 T pp and non-americans at 4 T pp.
http://www.anh-usa.org/wp-content/uploads/2016/04/ANHUSA-glyphosate-breakfast-study
[13] Peterson and Luxton, 1982; Lavelle and Spain, 2001
[14] DeAngelis 2016
[15] Wang, P., Wang, Y., & Wu, Q. S. (2016). Effects of soil tillage and planting grass on arbuscular mycorrhizal fungal propagules and soil properties in citrus orchards in southeast China. Soil and Tillage Research, 155, 54-61.
[16] Babikova, Z., Gilbert, L., Bruce, T., Dewhirst, S. Y., Pickett, J. A., & Johnson, D. (2014). Arbuscular mycorrhizal fungi and aphids interact by changing host plant quality and volatile emission. Functional Ecology, 28(2), 375-385.
[17] https://www.researchnester.com/reports/biofertilisers-market-global-demand-analysis-opportunity
[18] Zhu, X. G., Long, S. P., & Ort, D. R. (2008). What is the maximum efficiency with which photosynthesis can convert solar energy into biomass? Current opinion in biotechnology, 19(2)
[19] Morriën, E., Hannula, S. E., Snoek, L. B., Helmsing, N. R., Zweers, H., De Hollander, M. & Duyts, H. (2017). Soil networks become more connected and take up more carbon as nature restoration progresses. *Nature Comm*, *8*, 14349.
[20] Pers Comm with Dr. David Johnson, Adjunct Professor for the College of Agriculture at Chico State University.

Chapter Five

[21] New Zealand State of the Environment Report 2010.
[22]www.pesticideinfo.org/Detail_Poisoning.jsp?Rec_Id=PC3583
[23] Grasso, L. L., Martino, D. C., & Alduina, R. (2016). Production of Antibacterial Compounds from Actinomycetes. In *Actinobacteria-Basics and Biotechnological Applications* (pp. 177-198). IntechOpen.
[24] Personal note to Roger - beetroot it's a plant, not carrot-root or potato root!
[25] www.acsh.org/news/2018/07/28/geosmin-why-we-smell-air-after-storm
[26] Hemmings, S. M., Malan-Muller, S., van den Heuvel, L. L., Demmitt, B. A., Stanislawski, M. A., Smith, D. G., ... & Marotz, C. A. (2017). The microbiome in posttraumatic stress disorder and trauma-exposed controls: an exploratory study. *Psychosomatic medicine*, *79*(8), 936.
[27] Deurer, M., Grinev, D., Young, I., Clothier, B. E. and Müller, K. (2009), The impact of soil carbon management on soil macropore structure: a comparison of two apple orchard systems in New Zealand. European Journal of Soil Science, 60: 945–955.
[28] http://reefrescueresearch.com.au/news/183-pesticide-dynamics-in-the-gbr

[29] Negri Andrew P. , Flores Florita , Röthig Till , Uthicke Sven , (2011), Herbicides increase the vulnerability of corals to rising sea surface temperature, Limnology and Oceanography, 56,
[30] Chen, Y. & Aviad, T. 1990. Humic substances in soil and crop sciences: selected readings. Proceedings of a symposium cosponsored by the International Humic Substances Society, Chicago, Illinois.
[31] Pozdnyakov, I. P., Sherin, P. S., Salomatova, V. A., Parkhats, M. V., Grivin, V. P., Dzhagarov, B. M., ... & Plyusnin, V. F. (2017). Photooxidation of herbicide amitrole in the presence of fulvic acid. *Environmental Science and Pollution Research*, 1-8
[32] Erisman, Jan Willem; Sutton, Mark A.; Galloway, James; Klimont, Zbigniew; Winiwarter, Wilfried (2008-09-28). "How a century of ammonia synthesis changed the world". *Nature Geoscience*. **1** (10)
[33] http://www.pmcsa.org.nz/wp-content/uploads/PMCSA-Freshwater-Report.pdf
[34] Gourley, C. J., Dougherty, W., Aarons, S., & Kelly, K. Improving nitrogen use efficiency: from planet to dairy paddock
[35] Khan, S. A., Mulvaney, R. L., Ellsworth, T. R., & Boast, C. W. (2007). The myth of nitrogen fertilization for soil carbon sequestration. *Journal of Environmental Quality*, *36*(6), 1821-1832.
[36] Kuzyakov, Y., & Domanski, G. (2000). Carbon input by plants into the soil. Review. *Journal of Plant Nutrition and Soil Science*, *163*(4), 421-431.
[37] Bhardwaj, K. K. R. & Gaur A. C. The effect of humic and fulvic acids on the growth and efficiency of nitrogen fixation of *Azotobacter chroococcum.* Folia Microbiologica Volume 15, Number 5, 364-367.
[38] Poudel, D.D. Horwath, W.R. Mitchell J.P, & Temple, S.R. (2001) Impacts of cropping systems on soil nitrogen storage and loss. Agric. Syst., 68 (2001)
[39] Kramer, A. W., Doane, T. A., Horwath, W. R., & Kessel, C. V. (2002). Combining fertiliser and organic inputs to synchronize N supply in alternative cropping systems in California. *Agriculture, ecosystems & environment*, *91*(1), 233-243.
[40] Aguilera, E., Lassaletta, L., Sanz-Cobena, A., Garnier, J., & Vallejo, A. (2013). The potential of organic fertilisers and water management to reduce N2O emissions in Mediterranean climate cropping systems. A review. *Agriculture, Ecosystems & Environment*, *164*, 32-52.
41 Mouginot et al (2014) Elemental stoichiometry of Fungi and Bacteria from Grassland Leaf Litter. Soil Biology & Biochemistry 76 (2014) 278-285.
[42] Wang, G., Sheng, L., Zhao, D., Sheng, J., Wang, X., & Liao, H. (2016). Allocation of nitrogen and carbon is regulated by nodulation and mycorrhizal networks in soybean/maize intercropping system. *Frontiers in Plant Science*, *7*, 1901.
[43] Afkhami, M. E., & Stinchcombe, J. R. (2016). Multiple mutualist effects on genomewide expression in the tripartite association between Medicago truncatula, nitrogen-fixing bacteria and mycorrhizal fungi. *Molecular ecology*, *25*(19), 4946-4962.
[44] Griffiths, B. S. (1994). Microbial-feeding nematodes and protozoa in soil: Their effects on microbial activity and nitrogen mineralization in decomposition hotspots and the rhizosphere. *Plant and Soil*
[45] De Vries, F. T., Hoffland, E., van Eekeren, N., Brussaard, L., & Bloem, J. (2006). Fungal/bacterial ratios in grasslands with contrasting nitrogen management. *Soil Biology and Biochemistry*, *38*(8), 2092-2103
[46] De Vries, F. T et al (2012). Land use alters the resistance and resilience of soil food webs to drought. *Nature Climate Change*, *2*(4), 276-280
[47] Asghari HR, Cavagnaro TR (2012) Arbuscular mycorrhizas reduce nitrogen loss via leaching. *PLoSONE7*
[48] Pimentel, D., Culliney, T. W., Buttler, I. W., Reinemann, D. J., & Beckman, K. B. (1989). Low-input sustainable agriculture using ecological management practices. *Agriculture, ecosystems & environment*
[49] Chivenge, Pauline, Bernard Vanlauwe and Johan Six. "Does the combined application of organic and mineral nutrient sources influence maize productivity? A meta-analysis." *Plant and Soil* 342.1-2 (2011):
Chapter Seven
[50] Feynman, J. & Ruzmaikin, A. Climatic Change (2007) 84: 295).
Fu, C., & Dan, L. (2018). The variation of cloud amount and light rainy days under heavy pollution over South China during 1960–2009. *Environmental Science and Pollution Research*, *25*(3), 2369-2376.
Chambers, Jeffrey Q. and Paulo Artaxo. "Biosphere–atmosphere interactions: Deforestation size influences rainfall." *Nature Climate Change* 7.3 (2017): 175.

Evaluation on Cottonwood Ranch of Holistic Management. University of Nevada factsheet https://www.unce.unr.edu/publications/files/nr/2004/FS0467.pdf.
http://rri.ualberta.ca/Portals/115/Documents/Presentations/Teague_UofA_Feb15_ 2017.pdf?ver=2017-02-16-104508-157
Sanjari G, Ghadiri H, Ciesiolka CAA, Yu B (2008). "Comparing the effects of continuous and time-controlled grazing systems on soil characteristics in Southeast Queensland" Soil Research 46, 348–358.
Teague, R., Provenza, F., Kreuter, U., Steffens, T., & Barnes, M. (2013). Multi-paddock grazing on rangelands: Why the perceptual dichotomy between research results and rancher experience?. Journal of Environmental Management, 128, 699-717.
Arunkumar, N., Banu, J. G., Gopalakrishnan, N., & Prakash, A. H. (2018). *APPLICATION ASPECTS OF WAX DEGRADING BACTERIA FOR SUSTAINABLE CROP PRODUCTION*. EVERYMAN'S SCIENCE, 168.

[51] Ramanathan V, Crutzen PJ, Kiehl JT et al (2001) Aerosols, climate and the hydrological cycle. Science 294(5549):2119–2124

[52] Pielke, 2001

[53] Vick, E. S., Stoy, P. C., Tang, A. C., & Gerken, T. (2016). The surface-atmosphere exchange of carbon dioxide, water and sensible heat across a dryland wheat-fallow rotation. *Agriculture, ecosystems & environment*, *232*, 129-140.

[54] Taschetto, A. S., & England, M. H. (2009). An analysis of late twentieth century trends in Australian rainfall. International Journal of Climatology, 29(6)

[55] Failor, K. C., Schmale Iii, D. G., Vinatzer, B. A., & Monteil, C. L. (2017). Ice nucleation active bacteria in precipitation are genetically diverse and nucleate ice by employing different mechanisms. *The ISME journal*, *11*(12), 2740.

[56] Sheil, D. (2018). Forests, atmospheric water and an uncertain future: the new biology of the global water cycle. *Forest Ecosystems*, *5*(1), 1-22.

[57] Pokorný, Jan. "Dissipation of solar energy in landscape—controlled by management of water and vegetation." *Renewable energy* 24.3-4 (2001): 641-645.

[58] https://www.westernsustainabilityexchange.org/range-riders

[59] Kong, Dongdong, Heng-Cheng Hu, Eiji Okuma, Yuree Lee, Hui Sun Lee, Shintaro Munemasa, Daeshik Cho et al. "L-Met activates Arabidopsis GLR Ca2+ channels upstream of ROS production and regulates stomatal movement." *Cell reports*17, no. 10 (2016): 2553-2561.

[60] Tuason, M. M. S., & Arocena, J. M. (2009). Calcium oxalate biomineralization by Piloderma fallax in response to various levels of calcium and phosphorus. *Appl. Environ. Microbiol.*, *75*(22), 7079-7085.

[61] https://www.csiro.au/en/Research/AF/Areas/Sustainable-farming/Soil-water-landscape/Water-repellent-soils.

Chapter Eight

[62] Abbot, I. and C. A. Parker. (1981). Interactions between earthworms and their soil environment. Soil Biology and Biochemistry. 13, 191-197.
Basker, A., A. MacGregor and J. Kirkman. 1993. Exchangeable potassium and other cations in non-ingested soil and cast of two species of pasture earthworms. Soil Biology and Biochemistry. 25(12): 1673-1677.
Bohlen, P. and C. A. Edwards. 1995. Earthworm effects on N dynamics and soil respiration in microcosms receiving organic and inorganic nutrients. Soil Biology and Biochemistry. 27(3): 341-348.
Curry, J. P. and D. Byrne. 1992. The role of earthworms in straw decomposition and nitrogen turnover in arable land in Ireland. Soil Biology and Biochemistry. 24(12)
Edwards, C. A. 1995. Historical overview of vermicomposting. BioCycle. 36(6): 56-58.
Edwards, C. A. and J. E. Bates. 1992. The use of earthworms in environmental management. Soil Biology and Biochemistry. 14(12):1683-1689.
Elliot, P. W., D. Knight and J. M. Anderson. 1990. Denitrification in earthworm casts and soil from pastures under different fertiliser and drainage regimes. Soil Biology and Biochemistry. 22(5): 601-605.

Hopp, H. 1949. The effect of earthworms on the productivity of agricultural soil. Journal of Agricultural Research. 78(10): 325-339.
Joshi, N. V. and B. Kelkar. 1951. The role of earthworms in soil fertility. The Indian Journal of Agricultural Science. 22(2): 189-196.
Lee, K. 1985. Earthworms. Their Ecology and Relationships with Soil and Land Use. Academy Press
Nielson, R. 1965. Presence of plant growth substances in earthworms demonstrated by paper chromatography and the Went Pea Test. Nature. 208(5015):1113-1114.
Parkin, T. and E. Berry. 1994. Nitrogen transformations associated with earthworm casts. Soil Biology and Biochemistry. 26(9):1233-1238.
Ruz-Jerez, B., P. Roger and R. Tillman. 1992. Laboratory assessment of nutrient release from a pasture soil receiving grass or clover residues, in the presence or absence of Lumbricus rubellus or Eisenia fetida. Soil Biology and Biochemistry. 24(12)
Spain, A., P. Lavelle and A. Mariotti. 1992. Stimulation of plant growth by tropical earthworms. Soil Biology and Biochemistry. 16(2): 185-189.
Whitaker, T. and G. Davis. 1962. Cucurbits: Botany, Cultivation and Utilization. World Crops Books Interscience Publisher, Inc., New York.
Slade, E. M., Riutta, T., Roslin, T., & Tuomisto, H. L. (2016). The role of dung beetles in reducing greenhouse gas emissions from cattle farming. *Scientific reports*, *6*, 18140.
Losey, J. E., & Vaughan, M. (2006). The economic value of ecological services provided by insects. *AIBS Bulletin*, *56*(4), 311-323.
Grant, C., Bittman, S., Montreal, M., Plenchette, C., & Morel, C. (2005). Soil and fertiliser phosphorus: Effects on plant P supply and mycorrhizal development. *Canadian Journal of Plant Science*, *85*(1), 3-14.
Flaten, Sheppard 2001 phosphorus nutrition).
https://www.nasa.gov/nasa-satellite-reveals-how-much-saharan-dust-feeds-amazon-s-plants/
Bird excrement global contribution to nutrients: https://www.nature.com/articles/s41467-017-02446-8
Doty SL, Sher AW, Fleck ND, Khorasani M, Bumgarner RE, Khan Z, et al. (2016) Variable Nitrogen Fixation in Wild *Populus*. PLoS ONE 11(5): e0155979. https://doi.org/10.1371/journal.pone.0155979
Microbes synergistic for P efficiency: Baas, Peter; Bell, Colin; Mancini, Lauren M.; Lee, Melanie N.; Conant, Richard T.; Wallenstein, Matthew D. (2016). "Phosphorus mobilizing consortium Mammoth P™enhances plant growth". PeerJ. 4
Gonzalez, C. Y. C. J., Zheng, Y., & Lovatt, C. J. (2008). Properly timed foliar fertilization can and should result in a yield benefit and net increase in grower income. In VI International Symposium on Mineral Nutrition of Fruit Crops 868 . pp. 273-286
Ali, L.K.M. and Elbordiny M.M. (2009). Response of Wheat Plants to Potassium Humate Application.
Schefe, C. R. Et al. (2008). Organic amendment addition enhances phosphate fertiliser uptake and wheat growth in an acid soil.
Gale, D.L et al. (2011). Opportunity to increase phosphorus efficiency through co-application of organic amendments with mono-ammonium phosphate (MAP).
DORNEANU et al (2011). Efficacy of liquid organomineral fertiliser with humates extracted from lignite on leaf fertilisation of crops in the vegetation period.
Leach, K.A. and Hameleers, A. (2011). The effects of a foliar spray containing phosphorus and zinc on the development, composition and yield of forage maize.
Mosali, J. et al. (2006). Effect of Foliar Application of Phosphorus on Winter Wheat Grain Yield, Phosphorus Uptake and Use Efficiency. Potarzycki, J. and Grzebisz, A. (2009). Effect of zinc foliar application on grain yield of maize and its yielding components. S. H. Chien et al (2011). Agronomic and environmental aspects of phosphate fertilisers varying in source and solubility: an update review.
Hettiarachchi, G.M. et al. (2009). Reactions of Fluid and Granular Copper and Molybdenum-Enriched Compound Fertilisers in Acidic and Alkaline Soils. Schefe, C. R. Et al. (2008). Organic amendment addition enhances phosphate fertiliser uptake and wheat growth in an acid soil. Gale, D.L et al. (2011).
Gale, D. L. Opportunity to increase phosphorus efficiency through co-application of organic amendments with mono-ammonium phosphate (MAP). In *5 th World Congress on Conservation Ag*

[63] https://extension.psu.edu/earthworms
[64] Usmani, Z., Kumar, V., & Mritunjay, S. K. (2017). Vermicomposting of coal fly ash using epigeic and epi-endogeic earthworm species: nutrient dynamics and metal remediation. RSC Advances, 7(9),
[ii] Dey, M. D., Das, S., Kumar, R., Doley, R., Bhattacharya, S. S., & Mukhopadhyay, R. (2017). Ecological engineering, 106, 200-208.
[iii] Yadav, S. (2017). Potentiality of earthworms as Nanoscience and Plant–Soil Systems (pp. 259-278). Springer, Cham.
[65] Boots, B, Russell, C.W. and Green, D.S. 2019. Effects of Microplastics in Soil Ecosystems: Above and Below Ground. *Environmental Science & Technology* . September.
[66]Vijayabharathi R., Sathya A., Gopalakrishnan S. (2015) Plant Growth-Promoting Microbes from Herbal Vermicompost. In: Egamberdieva D., Shrivastava S., Varma A. (eds) Plant-Growth-Promoting Rhizobacteria (PGPR) and Medicinal Plants. Soil Biology, vol 42. Springer
[67] Brown, G.G., 1995. How do earthworms affect microfloral and faunal community diversity? Plant Soil
[68] Pringle, R. M., Doak, D. F., Brody, A. K., Jocqué, R., & Palmer, T. M. (2010). Spatial pattern enhances ecosystem functioning in an African savanna. PLoS biology, 8(5),
[69] Evans, T.A. et al. (2001) Ants and termites increase crop yield in a dry climate. Nat. Commun. 2:262 doi:0.1038/ncomms1257
[70] Kennedy Warne. Hotspot: New Zealand, National Geographic Magazine, October 2002.
[71] Bryony Sands et al. 2018. Sustained parasiticide use in cattle farming affects dung beetle functional assemblages , Agriculture, Ecosystems & Environment
[72] https://cen.acs.org/articles/95/i40/Drug-resistant-roundworms-prompt
[73] Errouissi, F., Alvinerie, M., Galtier, P., Kerboeuf, D., & Lumaret, J. P. (2001). The negative effects of the residues of ivermectin in cattle dung using a sustained-release bolus on *Aphodius constans* (Duft.)(Coleoptera: phodiidae). Veterinary Research, 32(5)
[74] Pat Coleby is the author of many books including "Natural Cattle Care" and "Natural farming." Her free choice mineral mix recipe is in these books.
[75] https://oregonstate.edu/instruct/css/330/three/Green.pdf.
[76] www.darrinqualmin.com
[77] https://ageconsearch.umn.edu/bitstream/183026/2/IAAE-CONF-222.pdf.
[78] Sparling, G.P. Shepherd, G.T. & Kettles, H.A. 1992 Changes in soil organic C, microbial C and aggregate stability under continuous maize and cereal cropping and after restoration to pasture in soils from the Manawatu region, New Zealand. Soil and Tillage Research, Volume 24, Issue 3,
[79] Daisog, H., Sbrana, C., Cristani, C., Moonen, A. C., Giovannetti, M., & Bàrberi, P. (2012). Arbuscular mycorrhizal fungi shift competitive relationships among crop and weed species. *Plant and soil*, *353*(1-2)
[80] Grant, C., Bittman, S., Montreal, M., Plenchette, C., & Morel, C. (2005). Soil and fertiliser phosphorus: Effects on plant P supply and mycorrhizal development. *Canadian Journal of Plant Science*, *85*(1), 3-14.
[81] Barkley, A. E. et al (2019) African biomass burning is a substantial source of phosphorus deposition to the Amazon, Tropical Atlantic Ocean and Southern Ocean. Proceedings of the National Academy of Sciences Jul 2019
[82]http://www.adfg.alaska.gov/index.cfm?adfg=wildlifenews.view_article
[83] Bird excrement global contribution to nutrients: www.nature.com/articles/s4146
[84] Baas, Peter; Bell, Colin; Mancini, Lauren M.; Lee, Melanie N.; Conant, Richard T.; Wallenstein, Matthew D. (2016). "Phosphorus mobilizing consortium Mammoth P™ enhances plant growth". PeerJ. 4

Chapter Nine

[85] Mahmood, A., Turgay, O. C., Farooq, M., & Hayat, R. (2016). Seed biopriming with plant growth promoting rhizobacteria: a review. *FEMS microbiology ecology*, *92*(8).

[86] Bidabadi, S. S., & Mehralian, M. (2019). Seed Bio-priming to Improve Germination, Seedling Growth and Essential Oil Yield of Dracocephalum Kotschyi Boiss, an Endangered Medicinal Plant in Iran. *Gesunde Pflanzen*, 1-11.

[87] White, J. F., Torres, M. S., Verma, S. K., Elmore, M. T., Kowalski, K. P., & Kingsley, K. L. (2019). Evidence for widespread microbivory of endophytic bacteria in roots of vascular plants through oxidative degradation in root cell periplasmic spaces. In *PGPR Amelioration in Sustainable Agriculture* (pp. 167-193).
[88] Journals of the Lewis and Clark Expedition. July 17, 1806

Chapter Eleven

[89] For a more in-depth look into what weeds are indicating read: "When weeds talk" by Jay McCaman.

[90] Matos, C. C., Costa, M. D., Silva, I. R., & Silva, A. A. (2019). Competitive Capacity and Rhizosphere Mineralization of Organic Matter During Weed-Soil Microbiota Interactions. *Planta Daninha, 37.*
[91] Trognitz, F., Hackl, E., Widhalm, S., & Sessitsch, A. (2016). The role of plant–microbiome interactions in weed establishment and control. *FEMS microbiology ecology, 92*(10).

[92] Samad, A., Trognitz, F., Compant, S., Antonielli, L., & Sessitsch, A. (2017). Shared and host-specific microbiome diversity and functioning of grapevine and accompanying weed plants. *Environmental microbiology, 19*(4), 1407-1424.
[93] Lei, S., Xu, X., Cheng, Z., Xiong, J., Ma, R., Zhang, L., ... & Tian, B. (2019). Analysis of the community composition and bacterial diversity of the rhizosphere microbiome across different plant taxa. *MicrobiologyOpen, 8*(6), e00762.

[94] Sindhu, S. S., Khandelwal, A., Phour, M., & Sehrawat, A. (2018). Bioherbicidal potential of rhizosphere microorganisms for ecofriendly weed management. In *Role of Rhizospheric Microbes in Soil* (pp. 331-376). Springer, Singapore.

[95] Elmore, M. T., White, J. F., Kingsley, K. L., Diehl, K. H., & Verma, S. K. (2019). Pantoea spp. Associated with Smooth Crabgrass (Digitaria ischaemum) Seed Inhibit Competitor Plant Species. *Microorganisms, 7*(5)
[96] Lawley, Y. E., J. R. Teasdale and R. R. Weil. 2012. "The Mechanism for Weed Suppression by a Forage Radish Cover Crop." Agronomy Journal 104: 205–214.
[97] https://www.slideshare.net/bio4climate/richard-teague-grazing-down-the-carbon-the-scientific-case-for-grassland-restoration-42538237.
[98] Schullehner, J., Hansen, B., Thygesen, M., Pedersen, C. B., & Sigsgaard, T. (2018). Nitrate in drinking water and colorectal cancer risk: A nationwide population-based cohort study. International journal of cancer, 143(1), 73-79.

Chapter Twelve

[99] Szczepaniec, A., Creary, S. F., Laskowski, K. L., Nyrop, J. P., & Raupp, M. J. (2011). Neonicotinoid insecticide imidacloprid causes outbreaks of spider mites on elm trees in urban landscapes. *PLoS One, 6*(5), e20018.
[100] Simon-Delso, N., Amaral-Rogers, V., Belzunces, L.P., Bonmatin, J.M., Chagnon, M., Downs, C., Furlan, L., Gibbons, D.W., Giorio, C., Girolami, V. and Goulson, D., 2015. Systemic insecticides (neonicotinoids and fipronil): trends, uses, mode of action and metabolites. *Environmental Science and Pollution Research, 22*(1),
[101] Mitchell, E. A., Mulhauser, B., Mulot, M., Mutabazi, A., Glauser, G., & Aebi, A. (2017). A worldwide survey of neonicotinoids in honey. *Science, 358*(6359),
[102] Pimentel, D. (1995). Amounts of pesticides reaching target pests: environmental impacts and ethics. *Journal of Agricultural and environmental Ethics, 8*(1), 17-29.
[103] M. Eng et al. A neonicotinoid insecticide reduces fueling and delays migration in songbirds. *Science.* Vol. 365, September 13, 2019.
[104] R.L Stanton, C.A. Morrissey and R.G. Clark. (2018) Analysis of trends and agricultural drivers of farmland bird declines in North America: A review. *Agriculture, Ecosystems & Environment.* Volume 254, February 15,
[105] Hageman, K. J., Aebig, C. H., Luong, K. H., Kaserzon, S. L., Wong, C. S., Reeks, T., ... & Matthaei, C. D. (2019). Current-use pesticides in New Zealand streams: Comparing results from grab samples and three types of passive samplers. *Environmental Pollution, 254,* 112973.

[106] Szczepaniec, A., Raupp, M. J., Parker, R. D., Kerns, D., & Eubanks, M. D. (2013). Neonicotinoid insecticides alter induced defenses and increase susceptibility to spider mites in distantly related crop plants. *PloS one*, *8*(5)
[107] Szczepaniec, Adrianna, Michael J. Raupp, Roy D. Parker, David Kerns and Micky D. Eubanks. (2013) "Neonicotinoid insecticides alter induced defenses and increase susceptibility to spider mites in distantly related crop plants." *PloS one* 8, no. 5 (2013)
[108] Chiriboga A (2009) Physiological responses of woody plants to imidacloprid formulations. MSc thesis. The Ohio State Univ. 130 p.
[109] https://practicalfarmers.org/wp-content/uploads/2018/10/Jonathan-Lundgren-Insects-and-Soil-Health.pdf
110 Gould, F., Brown, Z. S., & Kuzma, J. (2018). Wicked evolution: Can we address the sociobiological dilemma of pesticide resistance?. *Science*, *360*(6390), 728-732.
[111] Bass, C., & Jones, C. (2018). Editorial overview: Pests and resistance: Resistance to pesticides in arthropod crop pests and disease vectors: mechanisms, models and tools. *Current opinion in insect science*, *27*,
[112] https://www.epa.gov/sites/production/files/2014-10/documents/benefits_of_neonicotinoid_seed_treatments_to_soybean_production_
[113] Douglas, M. R., & Tooker, J. F. (2015). Large-scale deployment of seed treatments has driven rapid increase in use of neonicotinoid insecticides and preemptive pest management in US field crops. *Environmental science & technology*, *49*(8), 5088-5097.
[114] Bourguet, D., & Guillemaud, T. (2016). The hidden and external costs of pesticide use. In *Sustainable Agriculture Reviews* (pp. 35-120). Springer, Cham.
[115] LaCanne, C. E., & Lundgren, J. G. (2018). Regenerative agriculture: merging farming and natural resource conservation profitably. *PeerJ*, *6*, e4428.
[116] Thakur, M., Sohal, B. S., & Sandhu, P. S. (2016). Impact of elicitor spray on Alternaria blight severity and yield of Brassica juncea and Brassica napus species. Journal of Oilseed Brassica, 1(1), 78-82.
[117] Simon-Delso, N., Amaral-Rogers, V., Belzunces, L.P., Bonmatin, J.M., Chagnon, M., Downs, C., Furlan, L., Gibbons, D.W., Giorio, C., Girolami, V. and Goulson, D., 2015. Systemic insecticides (neonicotinoids and fipronil): trends, uses, mode of action and metabolites. *Environmental Science and Pollution Research*, *22*(1),
[118] Bohlen, P. J., & House, G. (2009). *Sustainable agroecosystem management: integrating ecology, economics and society*. CRC Press.
[119] Busch, J. W., & Phelan, P. L. (1999). Mixture models of soybean growth and herbivore performance in response to nitrogen–sulphur–phosphorous nutrient interactions. *Ecological Entomology*, *24*(2)
[120] Beanland, L., Phelan, P. L., & Salminen, S. (2003). Micronutrient interactions on soybean growth and the developmental performance of three insect herbivores. *Environmental Entomology*, *32*(3), 641-651.
[121] Tiwari, S., Singh, A., & Prasad, S. M. (2018). Regulation of Pesticide Stress on Metabolic Activities of Plant. In *Metabolic Adaptations in Plants During Abiotic Stress* (pp. 121-132). CRC Press.
[122] Sharma, E., Anand, G., & Kapoor, R. (2017). Terpenoids in plant and arbuscular mycorrhiza-reinforced defence against herbivorous insects. *Annals of botany*, *119*(5), 791-801.
[123] http://outgro.co.nz/wp-content/uploads/2013/10/Outgro-Report.pdf.
[124] Agnello, Art, Peter Jentsch, Elson Shield, Tony Testa and Melissa Keller. (2014) "Evaluation of Persistent Entomopathogenic Nematodes." Evaluation of Persistent Entomopathogenic Nematodes for Biological Control of Plum Curculio 22.1: 21-23. Cornell University Dept. of Entomology.
[125] An, R., Orellana, D., Phelan, L. P., Cañas, L., & Grewal, P. S. (2016). Entomopathogenic nematodes induce systemic resistance in tomato against Spodoptera exigua, Bemisia tabaci and Pseudomonas syringae. Biological control, 93,
[126]Kaaya, G. P., & Hedimbi, M. (2012). The use of entomopathogenic fungi, Beauveria bassiana and Metarhizium anisopliae, as bio-pesticides for tick control. *International Journal of Agricultural Sciences*, 2(6), 245-250.

[127] Boucias, D., Liu, S., Meagher, R., & Baniszewski, J. (2016). Fungal dimorphism in the entomopathogenic fungus Metarhizium rileyi: detection of an in vivo quorum-sensing system. Journal of invertebrate pathology, 136
[128] Vidal, S., & Jaber, L. R. (2015). Entomopathogenic fungi as endophytes: plant-endophyte-herbivore interactions and prospects for use in biological control. Curr Sci
[129] http://fantasticfungi.com/paul-stamets-bee-friendly/.
[130] Wei, Z., Gu, Y., Friman, V. P., Kowalchuk, G. A., Xu, Y., Shen, Q., & Jousset, A. (2019). Initial soil microbiome composition and functioning predetermine future plant health. *Science Advances*, *5*(9)
[131] Hale, A. N., Lapointe, L., & Kalisz, S. (2016). Invader disruption of belowground plant mutualisms reduces carbon acquisition and alters allocation patterns in a native forest herb. *New Phytologist*, *209*(2), 542-549.
[132] Mao, W., Schuler, M. A., & Berenbaum, M. R. (2017). Disruption of quercetin metabolism by fungicide affects energy production in honey bees (Apis mellifera). *Proceedings of the National Academy of Sciences*, *114*(10)
[133] Martínez-Medina, A., Fernández, I., Sánchez-Guzmán, M. J., Jung, S. C., Pascual, J. A., & Pozo, M. J. (2013). Deciphering the hormonal signalling network behind the systemic resistance induced by Trichoderma harzianum in tomato. *Frontiers in plant science*, *4*, 206.
[134] Worrall, D., Holroyd, G. H., Moore, J. P., Glowacz, M., Croft, P., Taylor, J. E., ... & Roberts, M. R. (2012). Treating seeds with activators of plant defence generates long-lasting priming of resistance to pests and pathogens. *New Phytologist*, *193*(3), 770-778.
[135] Dufour, R. (2006). Grapes: organic production. *United States: ATTRA-National Sustainable Agriculture Information Service.*
[136] Gatarayiha, M. C., Laing, M. D., & Miller, R. M. (2010). Combining applications of potassium silicate and Beauveria bassiana to four crops to control two spotted spider mite, Tetranychus urticae Koch. *International journal of pest management*, *56*(4),

Chapter Thirteen

[137] https://www.the-scientist.com/news-opinion/epa-cancels-registrations-for-12-neonicotinoid-pesticides-65956
[138] https://www.youtube.com/watch?v=4UkZAwKoCP8
[139] See Kate Indrelands General Mills profile on https://youtu.be/gWglTPo-FJk
[140] https://www.vbs.net.au/wp-content/uploads/2019/03/Graziers-with-better-profit-and-biodiversity_Final-2019.pdf
"Common Insecticide May Harm Boys' Brains More Than Girls". Scientific American. August 21, 2012.
U.S. EPA (2002). "Interim Reregistration Eligibility Decision for Chlorpyrifos"(PDF). November 19, 2012.

With Massive Gratitude

I never imagined that writing a book would be such a marathon, and at the same time, be so rewarding. None of this would've been possible without my family, in particular my brother, Jeremy. In the typical Masters family affair, he designed the book cover, read each chapter, offered critique, and has been there to film many of my soil adventures. He's the most knowledgeable city boy on the topic of soil. I love you so very much; I'm sorry for beating you up when we were kids.

I'm eternally grateful to my Mumsy, Michele, my best travel companion and biggest cheerleader. You pick me up and dust me off from any challenges and are there to celebrate the successes. You gave me fierce independence, my love for travel, and so many other personality quirks, that enable me to traverse through life with ease.

Dad, you are the kindest, most generous of spirits. You fostered my love for farming, nature and observation. This book would not have been possible without finance for my education, and the discovery of the "deceased worm farm estate." I am so grateful to not be saddled with student loan debt. I forgive you both for not getting me a puppy or a pony; I'm making up for it now.

Dear Nan-nan, you gave me solace, inspiration and many joy-filled memories; sitting on the carpet basking in the sunlight with the Encyclopedia Britannica and Louis Lamour for company. Thank you for your love, patience and rice puddings.

It was a community that helped me bring this book to fruition, the countless ears, patience, stories and critiques. And to those who have humoured one more story of "I'll be finished the book soon!" I'm forever indebted to Wendy Cashmore and Gwen Grelet, for their inspirational editorial help, astute insights and for checking my references. You are both kick-ass women, who provide leadership and inspiration to many; a beautiful demonstration that being involved in science and agriculture, does not mean sacrificing the essence of what it means to be a woman. Hats off to you my gorgeous friends!

To everyone in the Integrity Soils Team, Michael, Kim, Angus, Jules and Michelle, you indulge me as the 'star,' call me out on my BS and inspire me every day to transform my blind spots. You mostly tolerate my utter chaos with love, wit and some resignation that you never know where I am in the world. Thank you, Jules, for being my shoulder to cry on, and for holding us all as extraordinary. I'm honored to travel on this journey together with you in this amazing company and for "being" this work. You are all truly a demonstration of what 'Integrity' looks like in the world.

I have been fortunate to cross paths with some incredible women writers; Judith Schwartz, Didi Pershouse and Gretel Ehrlich. You provide huge inspiration to me. Thank you for your guidance and patience.

To my American family, Kate, Roger, Betsy and Anne, you are my port of call in any storm. You have embraced me in unconditional love, and you are as much family as any blood. Thank you for allowing my pony and I to park on your lawn as we pass through. Your morning coffee, hugs, pancakes, melted microwaves and enthusiasm for life and late-night brainstorms, provides me with a deep and rich well of motivation. Every day I count my blessings that our paths converged, and you were courageous to "just trial something," including friendship with a cocky Kiwi.

To all those who have all played a role in getting me here in one piece: Dave Pratt, Gretta Carney, Nick Pattison, Rachelle and Justin Armstrong, Caroline Masters, Aunty Cathy, Malou and the Anderson family, Marion Thompson, Bruce and Rachel Nimon, Tony Eprile, Monica Ravenheart, the WD crew, Paul and Elli Hawke, Katherine Cross, Charlotte Coddington and Mike Masters. To Wendy Millet, Elaine Patarini and all the Women in Ranching crew, my thunderbelly is over-flowing! And to the shrew-faced US customs officer, while slamming one door, you opened up an entire new world of opportunities, thank you.

To my mentors, enthusiasts and inoculators, who opened the door to the concepts and principles which underpin soil process: Dr Arden Anderson, Gary Zimmer, Jerry Brunetti, Di and Ian Haggerty, Steve Erickson, Betsy Ross, Graeme Sait, Daniel Hillel, Stuart Hill, Thelma and John Williams, Gary Zimmer, Masanobu Fukuoka and Elaine Ingham. And you Chrissy. And to all the regenerators, for your trust and love with inviting me into your homes, families and onto the land. You have been my greatest teachers, source of inspiration and the foundation that this book was built upon.

I want to thank EVERYONE who ever taught me something or said anything in encouragement. In particular a nostalgic thank you to Pennie Brownlee. I completed her 'Certificate in adult teaching' when I was 26. She began the program with the statement; "You cannot teach others, until you love yourself." It's been a long road, but now I see you were right, love is the basis for learning, curiosity and health. Pennie shared that that the most profound words ever spoken to her were "You are a master teacher, go out and teach." You passed that message on to me and I heard you! Thank you.

And to Bryn, who I dedicate this book to. Your generosity allowed me to follow my dreams, while you followed your path. Trust in this process; it is from digging deep that much gold lies and your strong warrior heart will shine. I love you to the moon and back.

About the Author

New Zealand born Nicole Masters, is an independent agroecologist, systems thinker, author and educator. She has a formal background in ecology, soil science and organizational learning studies in New Zealand. Nicole is recognized as a knowledgeable and dynamic speaker on the topic of soil health.

Her team of soil coaches at Integrity Soils, have a proven record working alongside food and fiber producers across the U.S., Canada, Australia and New Zealand, taking ag businesses to the next level in nutrient density, profitability and environmental outcomes.

Nicole has worked closely with a wide range of production sectors from; dairy, sheep & beef, viticulture, compost, nurseries, market gardens, racing studs, lifestyle blocks to large-scale cropping. Working alongside such diverse clients, has fostered a broad understanding of the challenges facing different production systems. She has devised and delivered educational programs for varied organisations; consultants, businesses, land care and extension services.

Nicole is one of a growing number of people who are facilitating the rapidly expanding world of quality food production and biological regenerative economies.

Nicole is currently based in North America, travelling with her horse and trailer. Her team at Integrity Soils are available for workshops, team training, facilitation, conferences and keynote presentations. Get in touch:

info@integritysoils.co.nz
www.integritysoils.co.nz

Made in United States
North Haven, CT
20 November 2022